James Webb

Reveals the Uniqueness of Telescope in Space

(The History of the World's Most Important Telescopes)

Harry Gleeson

Published By **Jenna Olsen**

Harry Gleeson

All Rights Reserved

James Webb: Reveals the Uniqueness of Telescope in Space (The History of the World's Most Important Telescopes)

ISBN 978-1-7781462-1-3

No part of this guidebook shall be reproduced in any form without permission in writing from the publisher except in the case of brief quotations embodied in critical articles or reviews.

Legal & Disclaimer

The information contained in this book is not designed to replace or take the place of any form of medicine or professional medical advice. The information in this book has been provided for educational & entertainment purposes only.

The information contained in this book has been compiled from sources deemed reliable, and it is accurate to the best of the Author's knowledge; however, the Author cannot guarantee its accuracy and validity and cannot be held liable for any errors or omissions. Changes are periodically made to this book. You must consult your doctor or get professional medical advice before using any of the suggested remedies, techniques, or information in this book.

Upon using the information contained in this book, you agree to hold harmless the Author from and against any damages, costs, and expenses, including any legal fees potentially resulting from the application of any of the information provided by this guide. This disclaimer applies to any damages or injury caused by the use and application, whether directly or indirectly, of any advice or information presented, whether for breach of contract, tort, negligence, personal injury, criminal intent, or under any other cause of action.

You agree to accept all risks of using the information presented inside this book. You need to consult a professional medical practitioner in order to ensure you are both able and healthy enough to participate in this program.

Table Of Contents

Chapter 1: What Is James Webb Telescope? 1

Chapter 2: Exoplanets Exploring New Worlds 17

Chapter 3: Webb Telescope in Discovering Explants 28

Chapter 4: Unveiling Earth-Like Explants 36

Chapter 5: The Importance of Finding Earth-Like 39

Chapter 6: Webb Telescope Can Help Become Aware About Habitable 42

Chapter 7: Future Prospects for Discovering Probably 48

Chapter 8: Unraveling the Mysteries of Stars 60

Chapter 9: The Capability of the James Webb Telescope 68

Chapter 10: The James Webb Space Telescope 73

Chapter 11: Sunshield Safety 82

Chapter 12: Optical Telescope Element .. 95

Chapter 13: Scientific Devices 108

Chapter 14: Spacecraft Bus 112

Chapter 15: Development and Statistics
.... 134

Chapter 16: The James Webb Space
Telescope 139

Chapter 17: The Science Goals 150

Chapter 18: The Data Processing and
Analysis 159

Chapter 19: The Role of Citizen Science 173

Chapter 1: What Is James Webb Telescope?

The James Webb Space Telescope (JWST) is a vicinity telescope especially constructed to undertake infrared astronomy. Its high-selection and excessive-sensitivity sensors allow it to study gadgets too historical, faraway, or dim for the Hubble Space Telescope. This lets in research spanning many domains of astronomy and cosmology, alongside detection of the number one stars and the start of the earliest galaxies, and thorough atmospheric characterization of in all likelihood livable explants.

The U.S. National Aeronautics and Space Administration (NASA) directed Webb's format and development and worked with key companies: the European Space Agency (ESA) and the Canadian Space Agency (CSA). The NASA Goddard Space Flight Center (GSFC) in Maryland supervised telescope development, on the identical time as the Space Telescope Science Institute in Baltimore on the Homewood Campus of Johns Hopkins University runs Webb. The important contractor for the challenge has emerged as Northrop Grumman. The telescope is known as after James E. Webb, who has become the administrator of NASA from 1961 to 1968 in the course of the Mercury, Gemini, and Apollo obligations.

The Webb became launched on 25 December 2021 on an Ariane 5 rocket from Kourou, French Guiana, and arrived at the Sun–Earth L2 Lagrange trouble in January 2022. The first Webb photograph becomes observed out to the sector in some unspecified time in the future of a information convention on 11 July 2022.

Webb's vital mirror comprises 18 hexagonal reflect segments built of antique-plated beryllium, which collectively make a 6.Five-meter-diameter (21 foot) reflect, in assessment with Hubble's 2.Four m (7 feet 10 in). This offers Webb a slight-collecting vicinity of spherical 25 square meters, or six instances that of Hubble. Unlike Hubble, which observes in the close to ultraviolet and visible (zero.1 to 0.8 μm), and near infrared (0.Eight–2.5 μm) spectra, Webb detects a decrease frequency variety, from prolonged-wavelength seen slight (crimson) to mid-infrared (zero.6–28.Three μm). The telescope want to be maintained very bloodless, under 50 K (−223 °C; −370 °F), in order that the infrared mild generated via the telescope itself does no longer intrude with the collected slight. It is placed in a sun orbit at the Sun–Earth L2 Lagrange factor, form of one.5 million kilometers (930,000 mi) from Earth, wherein its 5-layer sunshield shields it from warming with the aid of the Sun, Earth, and Moon.

Initial plans for the telescope, then titled the Next Generation Space Telescope, began in 1996. Two concept studies had been commissioned in 1999, for a prospective launch in 2007 and a US$1 billion expenditure. The software program become beset thru massive price overruns and delays.

A vast redesign was performed in 2005, with paintings completed in 2016 at a complete charge of US$10 billion. The excessive-stakes thing of the discharge and the telescope's complexity had been commented upon through the media, scientists, and engineers. In July 2023, scientists brought that the number one twelve months of James Webb Space Telescope operations emerge as a prime fulfillment.

The mass of the James Webb Space Telescope is type of half of that of the Hubble Space Telescope. Webb capabilities a 6.Five m (21 feet)-diameter gold-covered beryllium fundamental reflect made from 18 individual hexagonal mirrors. The replicate has a elegant ground of 26.Three m2 (283 squareToes), of which zero.Nine m2 (9.7 squareFeet) is hidden via the secondary resource struts,[16] presenting a entire collecting area of 25.Four m2 (273 sq.Feet). This is type of six instances extra than the collection location of Hubble's 2.4 m (7.Nine foot) diameter mirror, which has a meeting area of four.Zero m2 (forty 3 squareFt). The reflect has a gold coating to give infrared reflectivity and this is blanketed through a skinny layer of glass for sturdiness.

Webb is built specially for near-infrared astronomy, but can also view orange and crimson visible mild, similarly to the mid-infrared spectrum, relying at the machine being used. It can encounter items as a whole lot as one hundred times fainter than Hubble can, which embody subjects significantly in advance in the data of the universe, decrease again to redshift z≈20 (approximately one hundred eighty million years cosmic time after the Big Bang).For assessment, the first stars are considered to have formed among z30 and z≈20 (100–one hundred 80 million years cosmic time),[19] while the first galaxies may additionally have developed at redshift z≈15 (round 270 million years cosmic time). Hubble is not capable of view in addition once more than pretty early reionization at kind of z≈eleven.1 (galaxy GN-z11, 4 hundred million years of cosmic time).

The layout emphasizes the near to mid-infrared for severa reasons

immoderate-redshift (relatively early and far) gadgets have their seen emissions shifted into the infrared, and therefore their mild can be visible completely through infrared astronomy.Infrared light travels more without difficulty via dust clouds than visible mild. Colder gadgets which include particles disks and planets radiate most strongly inside the infrared;those infrared bands are hard to analyze from the floor or by means of using modern location observatories like Hubble.

Rough map displaying Earth's atmospheric absorption (or opacity) to distinct wavelengths of electromagnetic radiation, inclusive of seen moderate. Ground-based totally telescopes must see through Earth's surroundings, this is opaque in lots of infrared wavelengths (see photograph at right). Even even though the surroundings is apparent, most of the aim chemical additives, collectively with water, carbon dioxide, and methane, moreover exist in the Earth's environment, notably complicating studies. Existing space telescopes like Hubble can't test those bands as their mirrors are inadequately cold (the Hubble mirror is maintained at kind of 15 °C [288 K; 59 °F]) which means that that the telescope itself radiates intently in the applicable infrared bands.

Webb can also view devices in the Solar System at an mind-set of greater than 80 five° from the Sun and with an obvious angular charge of motion a whole lot less than 0.03 arc seconds in line with second. This contains Mars, Jupiter, Saturn, Uranus, Neptune, Pluto, their satellites, and comets, asteroids and minor planets at or past the orbit of Mars. Webb has the near-IR and mid-IR sensitivity if you want to stumble on nearly all appeared Kuiper Belt Objects In addition, it may study opportunistic and unplanned targets inside 48 hours of a desire to accomplish that, such as supernovae and gamma ray bursts.

Webb works in a halo orbit, revolving around a factor in place referred to as the Sun–Earth L2 Lagrange detail, extra or less 1,500,000 km (930,000 mi) beyond Earth's orbit throughout the Sun. Its actual location modifications amongst extra or less 250,000 and 832,000 km (one hundred fifty five,000–517,000 mi) from L2 as it revolves, preserving it out of every Earth and Moon's shadow. By way of reference, Hubble circles 550 km (340 km) above Earth's ground, whilst the Moon is around 400,000 km (250,000 mi) from Earth. Objects at this Sun–Earth L2 point can also circle the Sun in synchrony with the Earth, permitting the telescope to stay at a reasonably constant distance with continual orientation of its sunshield and system bus toward the Sun, Earth and Moon. Combined with its huge shadow-keeping off orbit, the telescope can concurrently block incoming heat and slight from all 3 of these our our bodies and avoid even the smallest changes of temperature from Earth and Moon shadows that might have an effect on the form, however however keep uninterrupted solar energy and Earth communications on its solar-going via aspect. This design keeps the temperature of the spacecraft regular and beneath the 50 K (−223 °C; −370 °F) vital for feeble infrared measurements.

James Webb Space Telescope sunshield

Test unit of the sunshield stacked and enlarged on the Northrop Grumman plant in California, 2014 To perform observations inside the infrared spectrum, Webb need to be maintained below 50 K (−223.2 °C; −369.7 °F); in any other case, infrared radiation from the telescope itself might overwhelm its tool. Its big sunshield protects moderate and heat from the Sun, Earth, and Moon, and its location close to the Sun–Earth L2 maintains all three bodies at the equal aspect of the spacecraft normally.Its halo orbit throughout the L2 element avoids the shadow of the Earth and Moon, keeping a constant environment for the sunshield and solar arrays.The consequent normal temperature for the systems at the dark aspect is crucial to making sure accurate alignment of the maximum essential reflect segments.

The 5-layer sunshield, every layer as skinny as a human hair is constructed of Kapton E movie, covered with aluminum on each aspects and a layer of doped silicon on the Sun-going thru issue of the 2 maximum updated layers to mirror the Sun's warmth decrease once more into location.Accidental rips of the sensitive movie framework throughout deployment sorting out in 2018 introduced about first rate delays to the telescope.

The sunshield grow to be envisage to be folded twelve instances (concertina style) in order that it'd in shape into the Ariane 5 rocket's payload fairing, that is 4.57 m (15.Zero ft) in diameter, and 16.19 m (fifty three.1 toes) lengthy. The shield's completely deployed dimensions have been projected as 14.162 m × 21.197 m (forty six.Forty six ft × 69.Fifty four ft)

Keeping below the color of the sunshield restricts the sphere of hobby of Webb at a person 2nd. The telescope can view 40 percent of the sky from someone problem, but can see all of the sky over a period of six months.

Telescope Element

Diffraction spikes way to reflect segments and spider colour-coded

Webb's important reflect is a 6.Five m (21 toes)-diameter gold-blanketed beryllium reflector with a gathering area of 25.4 m2 (273 squareToes). If it were built as a single, huge mirror, it would had been too brilliant for present day launch cars. The replicate is consequently fabricated from 18 hexagonal portions (a generation pioneered with the useful resource of Guido Horn d'Arturo), which unfurled after the telescope turned into launched. Image plane wavefront sensing using segment retrieval is accomplished to location the replicate segments inside the right spot the usage of precision actuators.

Subsequent to this initial setting, they in fact require rare upgrades each few days to keep splendid interest.This is in comparison to terrestrial observatories, for instance the Keck telescopes, which frequently regulate their replicate segments using active optics to counteract the results of gravity and wind loading.The Webb telescope makes use of 132 tiny actuation vehicles to put and adjust the lenses.The actuators can function the reflect with 10 nanometer precision.

Webb's optical format is a 3-reflect anastigmat,which takes use of curved secondary and tertiary mirrors to offer photographs which can be unfastened from optical aberrations during a huge vicinity. The secondary reflect is zero.Seventy 4 m (2.Four toes) in diameter. In addition, there is a nice guidance replicate that may range its role numerous times consistent with 2d to provide photo stabilization. Photographs acquired via Webb function six spikes plus fainter ones due to the spider assisting the secondary mirror.

Spacecraft bus (James Webb Space Telescope)

The spacecraft bus is the number one help factor of the James Webb Space Telescope, preserving a plethora of laptop, verbal exchange, electric powered powered energy, propulsion, and structural portions. Along with the sunshield, it constitutes the spacecraft part of the space telescope. The spacecraft bus is on the Sun-going thru "heat" aspect of the sunshield and runs at a temperature of about 3 hundred K (27 °C; 80 °F)

The framework of the spacecraft bus has a mass of 350 kg (770 lb), and need to help the 6,two hundred kilogram (thirteen,seven-hundred lb) area telescope. It is constructed normally of graphite composite cloth. The meeting turn out to be finished in California in 2015. It become blanketed with the relaxation of the gap telescope fundamental to its 2021 release. The spacecraft bus can rotate the telescope with a pointing precision of one arcsecond, and isolates vibration to 2 milliarcseconds.

Webb has pairs of rocket engines (one pair for redundancy) to make course corrections on the manner to L2 and for station maintaining – maintaining the correct characteristic in the halo orbit. Eight smaller thrusters are used for mind-set manipulate – the precise pointing of the spacecraft. The engines use hydrazine gas (159 liters or 40 U.S. Gallons at release) and dinitrogen tetroxide as oxidizer (79.Five liters or 21.Zero U.S. Gallons at release).

Webb isn't supposed to be serviced in area. A crewed challenge to restore or beautify the observatory, as emerge as finished for Hubble, must now not be viable, and steady with NASA Associate Administrator Thomas Zurbuchen, notwithstanding awesome efforts, an uncrewed a ways flung task changed into located to be past available generation on the time Webb grow to be designed. During the extended Webb finding out period, NASA officers noted the concept of a servicing challenge, however no plans had been announced.Since the a hit release, NASA has said that but limited lodging have become made to facilitate destiny servicing missions. These accommodations blanketed particular steering markers in the shape of crosses at the surface of Webb, to be utilized by far flung servicing missions, in addition to refillable fuel tanks, removable warm temperature protectors, and available attachment points.

NASA's James Webb Space Telescope is a real technological surprise. The biggest and most complex vicinity telescope ever built, Webb might be in a role to accumulate mild that has been visiting for thirteen.Five billion years, nearly thinking about that the start of the universe. In effect, Webb is a time system, permitting us to look at the number one galaxies to shape after the Big Bang. Because it gathers infrared mild, it will see proper thru the large clouds of dirt that block the view of maximum different telescopes. Webb can be one hundred times more powerful than the Hubble Space Telescope. Most appreciably, as quickly as it unfolds its 21-foot-huge (6.Five-meter-full-size) set of segmented mirrors, Webb is probably powerful sufficient to look for water vapor within the atmospheres of planets orbiting other stars.

Chapter 2: Exoplanets Exploring New Worlds

The James Webb Space Telescope (JWST) is an expected location-primarily based completely observatory slated to debut in 2021. It is a collaborative effort between NASA, the European Space Agency (ESA), and the Canadian Space Agency (CSA). This telescope is meant to be the successor to the Hubble Space Telescope and could offer unparalleled competencies to discover the cosmos.

One of the important thing scientific missions of the JWST is to analyze exoplanets, which might be planets past our sun device. With its cutting-edge device, the telescope can be capable of describe exoplanet atmospheres, take a look at their chemical compositions, and discover possible signs of habitability. By locating and studying exoplanets, scientists goal to get a higher know-how of ways planets amplify, evolve, and likely host life.

The telescope is prepared with 4 important scientific contraptions: the Near-Infrared Camera (NIRCam), the Near-Infrared Spectrograph (NIRSpec), the Mid-Infrared Instrument (MIRI), and the Fine Guidance Sensor/Near InfraRed Imager and Slitless Spectrograph (FGS/NIRISS). These device span a huge style of wavelengths, allowing astronomers to study celestial gadgets in numerous spectra, supplying a greater thorough comprehension of the cosmos.

The JWST's higher sensitivity and excessive-choice imaging capabilities will redesign our information inside the direction of numerous domain names of astronomy. For instance, it will allow researchers to have a take a look at the beginning region and development of galaxies greater successfully, which include the early cosmos. The telescope will accumulate slight from remote galaxies that has been stretched and redshifted thanks to the boom of location, revealing insights into the universe's infancy.

Additionally, the JWST can have a study the individual of dark depend and dark electricity, enigmatic components that make up a primary fraction of the cosmos but have not begun to be surely understood. By tracking the cosmic microwave historical past radiation and performing surveys of galaxy clusters, the telescope will help scientists gather important information on the ones unexplained occurrences.

Regarding predicted discoveries and breakthroughs, the JWST might in all likelihood likely display proof of the earliest galaxies that lengthy-installed after the Big Bang, presenting a view into the universe's beginnings. It may additionally moreover help solution essential issues regarding the origins of stars and planetary systems, offer notion at the atmospheres of exoplanets, and offer additional facts approximately the records of galaxies.

The telescope's discoveries may also have profound ramifications for our expertise of the cosmos. By investigating exoplanets, it would assist solution troubles regarding the likely predominance of lifestyles beyond Earth. By interpreting the riddles of darkish rely and dark strength, it's miles capable to reveal insights into the underlying nature of the universe and its very last destiny. Additionally, with the aid of monitoring distant galaxies, the JWST will assist to our knowledge of cosmic evolution and the beginning location of structures inside the universe.

It is vital to observe that the material provided right here is based totally on present day information as much as September 2023, and additional observe is needed to collect accurate and up to date records about the James Webb Space Telescope and its clinical

NASA's James Webb Space Telescope fixes its eyes on a preference of the greater than five,000 weird new planets determined out to exist in our galaxy, scientists will begin to map out a better picture of their "lives" - from start to loss of life.

So a protracted manner, scientists have determined that exoplanets - planets outside our solar device — exist in a significant range of sizes and bureaucracy. Webb, with its ability to reveal in no manner-before-seen factors of the cosmos the usage of infrared mild, will likely treatment a long time-lengthy disagreements over how planets originate and die.

From one million miles remote, the observatory will decide the composition of exoplanet atmospheres and explore their shape in 3 dimensions. And it would start to provide us a clearer photograph of planets like our personal — tiny, rocky, in all likelihood habitable worlds, and what it takes to supply them.

Discovery Alert: a 'Cool' Planet — with Plenty of Atmosphere?

Discovery Alert: A Giant Planet and Three Eclipses

"We're on the precipice of an explosion of our data of exoplanet atmospheres," stated Johanna Teske, a frame of humans scientist on the Carnegie Institution of Washington who's co-major a Webb statement institution with Natasha Batalha at NASA's Ames Research Center in Northern California.

"We'll recognise a few element extra than certainly their mass or period, and absolutely that they exist," Teske delivered. "We're beginning in case you need to pass the magnifying glass round."

Looking in on Planetary Nurseries, Youthful Planets

A key puzzle for astronomers: How do planets shape? Webb will utilize one in each of its maximum touchy sensors, the Mid-Infrared Instrument (MIRI), to discover the disks of gasoline and dirt whirling spherical newborn stars.

Scientists need to understand why those disks appear to have rings and gaps, says Charles Beichman, authorities director of the NASA Exoplanet Science Institute at Caltech and a high member in more than one Webb assertion missions. "Are planets within the technique of formation beginning these gaps?" he wonders.

Beichman and one-of-a-kind researchers also will probe for dusty leftovers of planet formation in faraway structures to recognize if they mimic our very own Kuiper Belt, the swarms of may-be comets in our outer sun system, or the asteroid belt between Mars and Jupiter.

A range of missions will target more youthful, immoderate-temperature planets, despite the fact that in the process of cooling and contracting after advent.

exoplanet spectrum

A transmission spectrum of a massive, warmness exoplanet, WASP-96 b, suggests the life of gases in its ecosystem, which encompass water vapor. The spectrum – from starlight filtered thru the planet's environment – turn out to be produced the use of the Webb telescope's Near-Infrared Imager and Slitless Spectrograph (NIRISS), and changed into some of the first clinical pics from Webb to be shared.

Using the MIRI tool and Webb's Near InfraRed Spectrograph (NIRSpec), the study company will collect spectroscopic measurements - splitting slight from the planets proper into a spectrum, setting up a form of fingerprint of chemical substances in those planets' atmospheres. That want to show houses like chemistry and the life of clouds, and supply essential insights to how those huge planets got here to be.

The Middle Years: Large and Small Planets

Planets at a mature diploma of improvement should in all likelihood inform us if the houses of planets in our non-public solar device –

likewise in their middle years – are common or unusual.

One Webb declaration organization hopes to have a observe the depths of a "heat Jupiter," HD 189733 b, which has been noticed through manner of previous satellite tv for laptop observatories. A little larger than our private Jupiter, this planet circles its superstar so carefully that a "year" takes truly 2 days.

The team, which includes Tiffany Kataria, an exoplanet scientist at NASA's Jet Propulsion Laboratory in Southern California, will take the ones earlier consequences to the following degree the usage of the MIRI tool. Researchers will perform spectroscopic measurements and "eclipse maps," recording atmospheric tendencies in all three dimensions even as the planet passes within the front of its movie star.

Mature planets embody smaller worlds, together with rocky planets in Earth's size range, and planets which are specifically huge, but however an entire lot a lot less than Neptune.

"We are interested in information the range and atmospheric compositions of planets between the scale of Earth and Neptune," Teske brought. "'Super-Earths' and 'mini-Neptunes' are the most not unusual forms of planets in our galaxy. But we handiest have some instances of atmospheric observations from those styles of planets."

Hovering inside the historical past is the regular issue matter of habitability, especially with terrific-Earths. "These planets which might be a chunk bit larger than Earth – can also need to they simply host habitable situations?" Teske asks.

Previous satellite television for computer telescope research have located rocky planets approximately Earth's period circling tiny, comparably chilly pink-dwarf stars.

Webb will seek for atmospheres in a brilliant planetary grouping named

TRAPPIST-1: seven about Earth-sized planets in tight orbits spherical a movie star a splendid deal lots less than 10 percentage the size of the Sun. One scientific software program will deal with

TRAPPIST-1e, which circles inside the midst of TRAPPIST-1's habitable region. Using NIRSpec, a crew headed with the beneficial useful resource of Nikole Lewis at Cornell University will motive to get spectroscopic measurements of the planet's surroundings — assuming it virtually has one.

Chapter 3: Webb Telescope in Discovering Explants

The James Webb Space Telescope (JWST) has the ability to play a large characteristic in coming across and analyzing exoplanets, which is probably planets positioned outdoor our sun device. Here are a few key techniques wherein the JWST can make contributions to exoplanet research:

Characterizing Exoplanet Atmospheres**: The JWST is ready with advanced spectroscopic gadgets, along with the Near Infrared Spectrograph (NIRSpec) and the Mid-Infrared Instrument (MIRI). These gadgets can examine the slight passing thru exoplanet atmospheres at some point of transits in the front in their host stars. By studying this "transit spectroscopy," scientists can encounter the presence of precise molecules and gases in exoplanet atmospheres, together with capacity symptoms of habitability or possibly biosignatures.

Studying Exoplanet Clouds and Weather**: The JWST's ability to have a study inside the infrared part of the spectrum lets in it to take a look at the weather patterns and cloud compositions of exoplanets. This can offer insights into the weather and atmospheric conditions on these distant worlds.

Characterizing Exoplanet Surfaces and Temperatures**: By looking at exoplanets inside the infrared, the JWST can assist estimate their ground temperatures and thermal emissions. This information is crucial for records whether or not or no longer these planets might also have conditions suitable for liquid water and, with the resource of extension, the potential for lifestyles.

Exploring Exoplanet Habitability**: JWST's observations can contribute to our know-how of exoplanet habitability by means of identifying planets with situations that would resource life as we're aware of it. This includes reading the presence of water vapor, carbon dioxide, and one-of-a-kind atmospheric components.

Detecting and Studying Exomoons**: JWST's immoderate sensitivity and precision can beneficial aid within the detection and characterization of exomoons, which orbit exoplanets. These moons might also moreover furthermore have their own atmospheres and could probably affect the habitability of their discern planets.

Investigating Exoplanet Formation**: The telescope also can contribute to our information of exoplanet formation by using way of watching protoplanetary disks spherical younger stars. This can help scientists hint the techniques that result in the transport of planetary systems.

Surveying a Wide Range of Exoplanets**: The JWST's huge problem of view and versatility make it able to gazing a good sized type of exoplanets, from gas giants to rocky terrestrial planets. This shape of desires allows for an entire exploration of the exoplanet population.

Providing Complementary Data**: The JWST's observations can complement statistics from different exoplanet-looking missions, which includes the Kepler Space Telescope and the Transiting Exoplanet Survey Satellite (TESS), via the usage of supplying first rate observe-up observations of pick exoplanets.

the James Webb Space Telescope is poised to revolutionize our data of exoplanets by way of manner of providing positive statistics approximately their atmospheres, compositions, and conditions. Its superior gadgets and abilities make it a vital tool for advancing the arena of exoplanetary technological understanding and the search for probably liveable worlds beyond our sun device.

Major exoplanet discoveries and their significance

key exoplanet discoveries and their relevance

1. Fifty one Pegasi b: Discovered in 1995, it became the primary examined exoplanet orbiting a primary-collection big name. Its discovery challenged prevailing perspectives on planetary formation and spread out the capability of discovering greater exoplanets.

2. Kepler-22b: Discovered with the useful resource of NASA's Kepler spacecraft in 2011, Kepler-22b modified into the primary exoplanet positioned in the liveable location of its movie star. It has a similar period to Earth and is a massive step in finding possibly liveable planets outside ourST-1 gadget: Discovered in 2016, the TRAPPIST-1 machine is a superb discovery. It includes seven Earth-sized planets, with three of them lying within the liveable location. This locating gives desire for the presence of probably habitable planets with activities favorable to life.

four. Proxima Centauri b: Discovered in 2016, Proxima Centauri b is the nearest acknowledged exoplanet to Earth. It circles the purple dwarf celebrity Proxima Centauri, it's miles part of the nearest big call system to us. Studying this planet may additionally moreover supply insights at the possibility for habitable habitats surrounding neighboring stars.

five. WASP-121b: Discovered in 2015, WASP-121b is an exoplanet orbiting incredibly near to its host big call, resulting in immoderate temperatures surpassing 2,500 ranges Celsius (4,500 levels Fahrenheit). By coming across this planet's ecosystem, scientists have emerge as considerable insights into the behavior of extremely-warmth gasoline giants.

These findings have drastically extended our records of exoplanetary systems and the opportunity for discovering liveable planets. They have furthermore advocated greater observe and technical inclinations in the location of exoplanet exploration.

How the telescope's advanced generation aids inside the take a look at of exoplanets

The telescope's superior tool advantages inside the studies of exoplanets in numerous methods. First, it permits astronomers to locate and describe exoplanets from extraordinary distances. Advanced telescopes can collect mild from those some distance flung planets, which permits estimate their period, composition, and unique vital elements.

Second, contemporary telescopes make use of severa strategies to investigate the atmospheres of exoplanets. By studying the moderate that travels thru or interacts with an exoplanet's atmosphere, scientists also can hit upon the life of severa molecules and factors. This offers significant knowledge approximately the planet's conditions and ability habitability.

Third, higher telescopes allow scientists to have a check exoplanets over extended intervals of time. This permits them to reveal modifications in the exoplanet's environment, weather patterns, and unique dynamic techniques. By analyzing these modifications, scientists may additionally acquire a greater facts of the planet's behavior and development.

Furthermore, current-day telescopes normally characteristic adaptive optics era, which allows accurate for atmospheric distortions and boom the readability of photos. This permits scientists to capture better and additional particular pix of exoplanets, giving greater information about their surfaces, structures, and ability moons.

Overall, the improved technology of telescopes drastically boosts our capacity to investigate exoplanets and enlarge our statistics of planetary systems past our very very own.

Chapter 4: Unveiling Earth-Like Explants

The quest for extraterrestrial lifestyles is of the exciting and thrilling disciplines of scientific exploration telescope era. Our capability to exoplanets has dramatically prolonged, developing the possibilities of probably livable worldsThe feasible outcomes of these for alien lifestyles encompass. Finding Earth-like exoplanets, those positioned in the liveable area in their large name in which times can be favorable for water and probable life, is a brilliant priority. The greater special and crisper pictures received using advanced telescopes permit scientists to acquire records on the atmospheres and compositions of these exoplanets.

By reading the environment exoplanet, scientists may additionally moreover hunt for proof of life, along with the life of particular gasses, consisting of oxygen, methane, or other natural compounds. These glasses might be consequences of natural sports activities sports taking area on the planet's surface. Additionally, a few molecules, inclusive of chlorophyll, might also recommend the life of plants.

Another effect is the probable finding of exoplan with moons. Moons play a key function with the aid of the use of way of stabilizing a planet's axis, generating thoughts, and giving extra resources of energy. Modern technology of telescopes lets in scientists to discover and examine the ones moons, our draw near of the complexity and of planetary structures.

Furthermore, the superior examine of exoplanets moreover permits scientists to higher preserve close to the times essential for their life. By evaluating the residences of numerous exoplanets, which includes period, composition, and distance their famous person, would possibly in all likelihood gather insights into the components that contribute to habitability.

Overall, a terrestrial life with the manner to study planets in extra element This extended records permits scientists to better understand probably livable planets and acquire proof of lifestyles beyond Earth. While the hunt for alien lifestyles stays a hard and project, those tendencies bypass us inside the route of addressing certainly in reality one of humanity's maximum critical questions

Criteria for an exoplanet to be taken into consideration Earth-like

Size and Composition: Ano Planet need to have a comparable size and composition to Earth. It want to be stony, in place of being greater regularly than not gas like Jupiter or Saturn.

It is important to cognizance on that those standards are based totally on our present information of habitability. As our knowledge and technology progress, the criteria also can trade or get greater present day

Chapter 5: The Importance of Finding Earth-Like

Finding Earth-like exoplanets is vital in the quest for life as it implies a probable gadget. Earth is the remarkable that sustains existence, consequently figuring out ex, composition, and distance from their host well-known individual enhances the possibility of locating possible homes for life.

If an exoplanet possesses Earth-like occasions, on the aspect of a robust floor, liquid water, and a sturdy environment, it might possibly the life of lifestyles as we're privy to it. These times are essential for the increase and sustenance of complex organisms Therefore, the finding of Earth-like exoplanets offers optimism to many varieties of lifestyles in the cosmos.

Furthermore, investigating Earth-like planets also can furthermore make bigger our information of the mechanisms concerned within the development and evolution of livable habitats. By the ones exoplanets to

Earth, scientists can also collect insights on the techniques that motive emergence and keeping life. This statistics may additionally moreover finally be hired to enhance searches, are searching for liveable exoplanets and beautify our opportunities of coming across alien life.

Additionally, the finding of Earth-like exoplanets creates hobby and curiosity inside the public, selling funding for location exploration scientific have a observe. It evokes our imagination and extends our view on the ability of existence beyond our personal planet.

However, it's miles essential to awareness on that coming across an Earth-like exoplanet does not endorse the existence of lifestyles. Many different tendencies, along side the planet's ecosystem, magnetic region, information, moreover play crucial roles identifying its habitability. Research and evaluation are required to reveal the life of existence on any exet, the locating of Earth conditions comes one step closer to addressing the query whether or not or no longer we're by myself in the cosmos.

Chapter 6: Webb Telescope Can Help Become Aware About Habitable

A rocky exoplanet orbiting a pink dwarf movie big name ninety eight light-years faraway might also include the vital detail of the manner in all likelihood it's far for planets like Earth to show into uninhabitable worlds like Venus.

The exoplanet, referred to as LP 890-9c (additionally called SPECULOOS-2c), became decided in September 2022. It has a diameter 40% huge than Earth's and circles its movie star every eight.5 Earth days at a distance of best 1.7 million miles (2.Eight million kilometers). The pink dwarf, nevertheless, is tiny and cold, which means that temperatures may be smooth even close to to the celeb. Indeed, LP 890-9c is positioned closer to the internal boundary of the megastar's liveable area, it certainly is the distance round a film megastar in which a planet with an Earth-like surroundings might possibly preserve liquid water on its surface.

New have a have a look at performed with the aid of Lisa Kaltenegger, director of the Carl Sagan Institute at Cornell University, has analyzed the capability climatic and atmospheric states that LP 890-9c can be in, and the way the James Webb Space Telescope ought to figure amongst them.

Lp 890-9c's feature in its planetary device is comparable to Venus' placement in our solar device, which is likewise close to the inner boundary of the habitable place. In idea, a planet at Venus' function can also moreover stay liveable, but in a few unspecified time within the destiny in its 4.5-billion-one year records, Venus have become locked within the comments loop of a runaway greenhouse effect. Any water Venus formerly had on its floor boiled away, and the planet was left with a thick carbon-dioxide environment.

However, not all planets at the inner border of the habitable place will usually increase the identical way that Venus has. For one, Venus does now not have its very personal magnetic scenario to fend off the sun wind, the torrent of charged debris pouring from the sun. This made it much less difficult for the solar wind to transport away hydrogen atoms that have been damaged from water molecules with the useful resource of ultraviolet radiation from the sun, thereby depriving the planet of water. If LP 890-9c has a sturdy magnetic issue, it may be able to combat off its massive name's stellar wind and keep onto the water vapor in its surroundings.

"Looking at this planet will tell us what's occurring on the inner fringe of the habitable sector — how lengthy a rocky planet can preserve habitability at the equal time as it begins to get warm," Kaltenegger said in a launch.

Kaltenegger's organization modeled the planet based on its mass and radius, every of which astronomers have measured. The models additionally protected assumptions approximately the planet's chemical composition, its floor temperature and stress, the intensity of its surroundings and the quantity of cloud gift. These latter elements are but doubtful. Indeed, the planet may be airless and cratered for all we understand, this is a sizable possibility for the motive that pink dwarfs are generally susceptible to excessive flares that could take an environment from a close-by orbiting globe.

The researchers created five capability fashions depicting what LP 890-9c might be like. These numerous from an Earth-like planet however hotter, via varying levels of atmospheric water vapor interest and stages of greenhouse outcomes, completing within the very last version comparable to hellish Venus with its choking carbon dioxide.

A separate observe, led with the resource of Jonathan Gomez Barrientos of the California Institute of Technology, shows that JWST may want to first-class need to take a look at three transits of LP 890-9c across the face of its host celebrity to verify a steamy, water-wealthy surroundings; eight transits to acquire enough information to decide whether or not LP 890-9c is extra like Venus; and 20 transits to discover proof of the despite the fact that-liveable warmth Earth state of affairs. Since the planet transits its huge name each eight.Five Earth days, the observations may additionally in idea take most effective six months to make.

"This planet is the number one aim in which we will check the ones one-of-a-kind scenarios," stated Kaltenegger. "If it is though a hotter Earth — warm, but with liquid water and conditions for life — then the internal edge of the habitable place [around all stars] can be teeming with lifestyles."

JWST cannot at once come upon water inside the world's floor, however it might establish if the surroundings's composition may additionally want to in form the lifestyles of liquid water.

Even if LP 890-9c appears to be too warm for lifestyles, the facts would in all likelihood tell us about the future of our non-public planet,

Earth. As the sun matures, it will slowly get brighter and hotter; after greater or less 1000 million years' time it will have rendered Earth too warmth for existence to thrive, and the seas will evaporate. Studying a planet aside from Venus that has already long beyond through this phase, or likely even resisted it for now, can assist teach us about Earth's very personal destiny.

The James Webb Space Telescope (J is an advanced region this is meant for our potential to test ex atmospheres. Here are some approaches in which the JWST ought to in all likelihood assist find liveable exoplanets:

Chapter 7: Future Prospects for Discovering Probably

This is a extraordinarily thrilling period for exoplanet research. With the chronic growth in the amount and form of exoplanets there may be honestly loads to be learnt and, as we circulate from an age of locating exoplanets to taken into consideration one in each of characterizing them, entire new fields of inquiry may in all likelihood appear. For example, with the resource of similarity with our Solar System, moons want to exist surrounding exoplanets, and people can also moreover offer feasible houses for lifestyles. The hunt for exomoons is continuing, however none had been discovered to this point.

A lot of what we recognize approximately exoplanets – even excessive examples which embody the hypothetical diamond planets and lava worlds – are though genuinely interesting hints that require extra complete and specific measurements in advance than some element may be verified. The CoRoT and Kepler area telescopes transformed the check of exoplanets, and we might also live up for an interesting time earlier with the future place missions. Building on what has been learnt so far, the brand new missions are supposed to recognize tiny planets surrounding tremendous stars. Because the host stars are high-quality the masses of the diagnosed planets can be calculated from radial pace facts at floor-primarily based absolutely observatories — both mass and length are crucial for characterizing the modern-day well-knownshows.

TESS

In 2018, NASA launched TESS (Transiting Exoplanet Survey Satellite) to do an all-sky survey to display notable stars for tiny transiting planets. TESS stars are 30–one hundred instances brighter than the ones studied with the aid of way of the Kepler spacecraft, which means that that TESS planets need to be significantly less tough to define with observe-up studies. These comply with-up observations will provide precise measurements of the planet hundreds, diameters, densities, and atmospheric parameters. Planned to carry out for two years, TESS will provide properly goals for added, extra thorough description with future massive floor-primarily based and place-primarily based completely observatories.

CHEOPS Europe's feature ExOPlanets Satellite (CHEOPS) venture, a cooperation amongst ESA and Switzerland, come to be launched in December 2019 and started operations in April 2020. Using the method of immoderate accuracy transit photometry, this challenge investigates diagnosed exoplanets which might be smaller than Saturn and are circling close to to vibrant stars. A vital characteristic of CHEOPS is its comply with-up nature: it video show gadgets particular stars acknowledged to residence planets, instead of sporting out sky surveys to look for greater. Since scientists will understand precisely even as and wherein to element the satellite tv for laptop to seize the exoplanet as it transits the disc of its host famous character, it'll likely be viable to have a take a look at a couple of planetary transits to build up immoderate precision at the shallow transit signatures of the smaller planets inside the Earth–Neptune length variety.

By combining correct and particular measurements of an exoplanet's radius – received with CHEOPS – with present mass determinations, scientists can determine accurate densities for a large sample of planets, as a result making the number one steps inside the characterisation of these exoplanets and placing constraints at the composition of these smaller planets.

Understanding the real nature of planets desires no longer satisfactory measures of mass and radius, however additionally a take a look at of their atmospheric skills. By figuring out the bodily amount of the atmospheres of awesome-Earths, CHEOPS may be able to discriminate among Earth-like planets wherein lifestyles as we apprehend it may blossom and top notch kinds of Earth-mass planets (hydrogen-rich Earths, ocean planets), which mission our modern-day information of habitability. As such, CHEOPS will offer unique desires for the James Webb Space Telescope similarly to for the subsequent generation of ground-based totally completely very massive telescopes, both of with a motive so one can analyzing the fingerprints of molecules in the atmospheres of neighboring exoplanets.

CHEOPS may additionally even measure the slight curves for a small pattern of warm-Jupiters and find out the strategies via which energy is carried through and across the exoplanet ecosystem. As with any medical assignment, CHEOPS might be utilized to remedy technological knowledge problems which is probably extra speculative, such as looking for the photometric signatures of exomoons (moons orbiting exoplanets), earrings and Trojans, and will also examine topics outside exoplanet research.

WEBB

The NASA/ESA/CSA James Webb Space Telescope, slated to debut in 2021, will deliver project-changing new competencies for investigations of exoplanets and their atmospheres. With a collection of four sensors strolling at infrared wavelengths Webb will look at a couple of techniques to find out those extrasolar planets.

Highly sensitive spectroscopic assertion of transiting planets – with comparable talents in terms of length and mass – will bring in within the age of comparative planetology for exoplanets. Webb will take a look at exoplanet atmospheres thru taking photographs absorption, reflected picture, and emission spectra at infrared wavelengths for planets masking numerous sizes, from first rate-Earths to fuel giants. It will take benefit of the reality that, at those wavelengths, molecules in the atmospheres of exoplanets display a massive shape of spectral traits, giving the observers a entire set of diagnostic device, a whole lot of which are not to be had from the ground.

Webb also can be capable of straight away picture positive young and big exoplanets orbiting at wider distances from their determine big name than most transiting ones. Three of Webb's sensors embody immoderate-evaluation imaging skills (in two instances that is executed through a coronagraph) to lessen the glare of the parent celeb, making it much less hard to picture the planet. Observations with brilliant infrared filters will provide some of facts approximately the ones planets, their traits and their creation strategies.

PLATO

ESA's PLATO (PLAnetary Transits and Oscillations of stars) challenge is slated to release in 2026. PLATO is bear in mind to discover and look at a notable extensive range of new extrasolar planetary systems, with the aid of the usage of reading loads of hundreds of awesome stars for transiting planets. PLATO can have the particular capability to discover and determine the tendencies of terrestrial planets that orbit inside the habitable zone round stars just like our Sun.

By combining the PLATO accurate measurements of radii for a huge sample of planets with the respective planetary loads determined from ground-primarily based truly observations, scientists may be able to find out the diversity of planets that exist, which in flip offers critical constraints on how planets shape. These studies might also even permit scientists to estimate the mass composition of a massive quantity of youth planets, to evaluate how similar they may be to the Earth, and to assess their habitability.

How the host superstar impacts the planets which can be in orbit round it is an vital trouble that PLATO information is probably used to investigate. Only by using manner of reading the host movie star attributes, along facet stellar interest, kind, metallicity, and so on., can we apprehend planetary systems. For the number one time, PLATO will allow scientists to compute precisely the parameters of a huge type of stars with planets, which include their a long term. With this information, it's going to likely be able to examine the adjustments that planets experience with time, and to apprehend how the situations for habitability increase.

The large pattern of systems if you want to be decided with the useful resource of PLATO may additionally even deliver information to research the structure and development of extrasolar planetary structures, by the use of manner of analyzing the distribution of planet types – terrestrial or gaseous – with distance from the host famous individual and with stellar age. This will assist scientists to set our Solar System in attitude - is its makeup normal or unusual?

By detecting planets surrounding colorful stars, PLATO may be a pathfinder for subsequent missions searching out warning signs and symptoms of existence — the ones types of planets are the best candidates for spectroscopic have a look at-up observations to assess the structure and composition of planetary atmospheres.

ARIEL

Moving in advance past locating inside the route of investigating and statistics, ESA's Ariel (Atmospheric Remote-sensing Infrared Exoplanet Large-survey) assignment will adopt a chemical census of a massive, properly-described and varied pattern of exoplanets. Conducting simultaneous observations in a few unspecified time in the future of a number of visible and infrared wavelengths, the venture will allow a have a observe of exoplanets each as humans and as businesses.

Due to release in 2029, Ariel is supposed to undertake immoderate-accuracy transit, eclipse, and segment-curve multiband observations making use of simultaneous photometry in the seen and spectroscopy in close to infrared wavelengths. It will come across and have a look at about one thousand preferentially warmth and hot transiting gas giants, Neptunes, and tremendous-Earths round some of well-known character kinds and planetary device designs.

Ariel will produce an first rate database of planetary spectra, characterizing molecular abundances, chemical gradients, atmospheric structure, diurnal and seasonal fluctuations, clouds, and albedo measurements. The undertaking seeks to offer a totally representative picture of the chemical make-up of the exoplanets investigated, and additionally tie this proper away to the kind and chemical composition in their host stars, permitting scientists to find out the character of these planets, how they originated and the way they boom.

With this array of space telescopes arriving over the subsequent decade, we might also moreover furthermore expect getting in the direction of locating "Earth 2.0", on the equal time as on the identical time inclusive of extra unusual and surprising planets to the exoplanetary zoo. Exciting moments lay coming.

Chapter 8: Unraveling the Mysteries of Stars

As we maintain to push the frontiers of area studies, the James Webb Space Telescope (JWST) emerges as an brilliant device that guarantees to unveil the mysteries of Orion's seven distinguished stars and supply crucial insights into the universe's beginnings and improvement.

Behold the Splendor of Orion: An innovative image showcasing the stars of the Orion constellation accompanied with a drawing inspired with the resource of the historic Greek photo of the heroic hunter. Credit: NASA/STScI Aug 07, 2023 - The expanse of vicinity has prolonged concerned human creativeness, and some of the most awe-inspiring celestial beauties is the Orion constellation, typically known as "The Hunter." This beautiful affiliation of stars has attracted astronomers for a long term, with its precise belt and a superb employer of neighboring stars. As we maintain to push the frontiers of place research, the James Webb

Space Telescope (JWST) emerges as an extraordinary device that promises to unveil the mysteries of Orion's seven excellent stars and deliver critical insights into the universe's beginnings and improvement.

The Celestial Masterpiece of Orion's Seven Prominent Stars

Orion rises tall as one of the most proper now identifiable constellations in each the Northern and Southern Hemispheres. Dominating the wintry weather night time sky, this conventional constellation is a celestial masterwork that has led explorers and dreamers for many years. Orion is decorated with numerous vibrant stars, each bearing its unique attraction and significance.

The seven wonderful stars in the Orion constellation consist of:

Betelgeuse (Alpha Orionis): Betelgeuse is a crimson supergiant and one of the brightest stars within the night time sky. Located in the shoulder of Orion, this massive superstar is almost seven hundred times larger than our Sun and is diagnosed for its erratic fluctuation, periodically dimming and brightening through the years.

Rigel (Beta Orionis): Rigel is a blue-white supergiant and one of the brightest stars in the sky. Positioned near Orion's knee, it outshines Betelgeuse and ranks because the seventh brightest big name within the night time sky. Rigel's brightness makes it a tremendous aim for statement with the JWST.

Bellatrix (Gamma Orionis): Found near Orion's left shoulder, Bellatrix is a large blue famous person that contributes to the constellation's splendor. Its name interprets to "Female Warrior" in Latin, perfectly representing its place as a part of Orion's discern.

Mintaka (Delta Orionis): As one among Orion's belt stars, Mintaka office work a straight away line with Alnilam and Alnitak, identifying the Hunter in the night time time time sky.

NAlnilam (Epsilon Orionis): Also called "Al Nihal," Alnilam is every other one in each of Orion's belt stars, placed in step with Mintaka and Alnitak.

Alnitak (Zeta Orionis): The zero.33 of Orion's belt stars, Alnitak completes the distinctive line that paperwork Orion's belt.

Saiph (Kappa Orionis): Located close to Orion's proper knee, Saiph is a blue-white supergiant, and on the identical time as not as well-known because the opportunity stars, it contributes to the constellation's luminosity.

The James Webb Space Telescope: Unleashing Unprecedented Power

The James Webb Space Telescope marks the apex of area observatories, a joint business enterprise amongst NASA, the European Space Agency (ESA), and the Canadian Space Agency (CSA). This new telescope has leap forward technology that transcends its predecessors, which includes the Hubble Space Telescope.

a. Infrared Capabilities: Operating predominantly within the infrared spectrum, JWST boasts the best benefit of seeing via cosmic dust clouds, permitting a sharper photograph of faraway stars and galaxies, which consist of the ones in the Orion constellation. Its higher infrared sensitivity exposes fainter and colder celestial objects formerly undetected via traditional telescopes.

b. Large Mirror: JWST possesses a large 6.Five-meter segmented mirror, dwarfing the Hubble's. This large replicate collects greater mild, giving in crisper pictures and heightened sensitivity, top notch for viewing faint stars similar to those indoors Orion's lap.

c. State-of-the-art Instrumentation: Equipped with a set of modern day medical system, JWST allows astronomers to investigate well-known person formation, the makeup of atmospheres, and the homes of exoplanets with unheard of accuracy. These gear promise to expand our take a look at of the celebrities internal Orion's dominion.

Unveiling the Enigmatic Stars of Orion

The aggregate of Orion's closeness to Earth and the JWST's competencies gives a big quantity of have a examine options for astronomers:

a. Stellar Birth: Orion is a middle of stellar nurseries, diagnosed for its active movie star-forming areas. JWST's infrared imaginative and prescient permits scientists to penetrate thick dust and gasoline surrounding those stellar nurseries, revealing perspectives of the early levels of celebrity formation and unveiling the secrets and strategies of stellar begin.

b. Stellar improvement: The take a look at of Orion's notable stars will extend our information of stellar development, from extra youthful protostars to mature giants. JWST's infrared observations will permit astronomers to discover essential developmental ranges, providing a entire image of well-known individual lifestyles cycles.

c. Exoplanets: JWST's robust spectroscopic gadgets will have a look at the atmospheres of exoplanets interior Orion's neighborhood. This new capability may screen viable habitability and offer information about the presence of alien existence.

d. Dark Clouds and Nebulae: Orion's constellation includes captivating darkish clouds and nebulae. JWST's infrared imaginative and prescient will untangle their chemical compositions, uncovering the mysterious situations wherein stars and planets are created.

Orion's seven brightest stars have attracted human beings thinking about that historical times, generating hobby and awe approximately the sky. The James Webb Space Telescope marks a brand new era in place research, allowing scientists to dive deeper into the riddle of Orion's stars and garner fundamental insights into the universe's beginnings and development. As JWST starts off evolved off evolved on its awe-inspiring voyage, the scene is ready for soar ahead discoveries that permit you to remodel our records of the universe and our function interior it. Through this first rate partnership a few of the JWST and the stars of Orion, mankind is prepared to get to the lowest of the mysteries of the cosmos and circulate even within the route of the frontiers of information.

Chapter 9: The Capability of the James Webb Telescope

The James Webb Space Telescope, the prolonged awaited, groundbreaking entrant to the pantheon of area observatories, is now approximately to make a superb splash, a nice departure from the waves its oft-behind schedule launch and growing price have brought approximately. After many hold u.S. Of americathis 12 months, NASA is making plans to launch the craft on Christmas. The telescope's launch has now been not on time a decade, and its price has elevated approximately $nine billion above finances. Lawmakers and scientists each have voiced fear that the mission is siphoning funding from unique medical areas, at the same time as many precise experts feel that Webb is in reality nicely well worth the coins and the wait.

Webb's format is inspired through the Hubble Space Telescope, the 31-three hundred and sixty five days-antique telescope famed for generating fantastic photos of our universe's galaxies. But Webb takes up in which its predecessor falls brief, says Eric Smith, Webb's software scientist and head scientist of NASA's Astrophysics Division. There's in fact no telescope like Webb so far, he gives. The new observatory, which is prepared to release from northern French Guiana close to the equator, is a collaborative mission related to the space companies of the united states, Europe and Canada. "When you notice Webb pass into vicinity, ... It's the complete pressure of human creativity and all kinds of disciplines that push it there."

This maximum contemporary area-certain gizmo is unique because of traits. First, it's big, with a 21.Three-foot vital mirror as a way to make Webb the farthest-seeing telescope humanity has ever created. Secondly, Webb perceives the cosmos inside the infrared—the vicinity on the electromagnetic spectrum with simply longer wavelengths than seen light. It might be the number one infrared-specialized telescope in orbit that might have a study high-quality distances. Its nearest opponent,

Hubble, operates thru and large within the visible and has a confined infrared-viewing range.

Below are five subjects the telescope will allow astronomers to perform.

Understand how early galaxies started out and superior

The Rubin Galaxy, named for the astronomer Vera Rubin, twirls in area 232 million slight years far off from Earth. NASA, ESA, and B. Holwerda, University of Louisville

"One of the big functions of telescopes is sincerely as time machines, because distance is look-once more time," explains Daniel Eisenstein, an astronomer at the Harvard–Smithsonian Center for Astrophysics. Eisenstein will use Webb's cameras to "time-tour" returned to at the identical time because the primary galaxies have been formed proper away after the Big Bang.

When we gaze at a far off galaxy mild years away, we aren't viewing it in its most cutting-edge situation. Its distance in slight years equates to the amount of years it takes for its mild to gain Earth. For instance, the closest galaxy to ours is the Canis Major Dwarf Galaxy this is 25,000 slight-years remote, therefore its moderate takes 25,000 years to achieve Earth. That way even as we have a take a look at Canis Major Dwarf, we're seeing it as it emerge as 25,000 years in the beyond.

The deeper into region astronomers can gaze, the similarly returned in time they might examine a galaxy. Webb, being the farthest seeing telescope ever, can choose out the youngest appearing galaxies people can encounter. To apprehend the advent of galaxies, scientists like Eisenstein will examine multiple galaxies at numerous existence degrees and put together their growth chronology.

Webb's infrared abilities also are vital for viewing the ones galaxies. Light from far off galaxies may be stretched out by manner of the use of the increasing cosmos. By the time the moderate reaches our telescopes, its unique wavelength may have moved from the visible or ultraviolet to the infrared. Luckily, choosing up infrared signs is surely up Webb's alley. "It's the primary time we've got had a big, cold telescope in area that could examine those infrared wavelengths," gives Eisenstein.

The Hubble vicinity telescope has succeeded to gather the shortest wavelength infrared rays prolonged from the bluest of mild of remote galaxies. The decommissioned Spitzer infrared telescope have end up a long way smaller than Webb and couldn't view as a long manner into vicinity. Webb will hit it out of the park in phrases of methods a long manner into place—and the manner a long manner lower again in time—it is able to seize some distance off galaxies in the gadget of growing up.

Chapter 10: The James Webb Space Telescope

The James Webb Space Telescope (JWST) is an area telescope presently accomplishing infrared astronomy. As the maximum crucial optical telescope in location, it's miles organized with excessive-decision and immoderate-sensitivity gadgets, allowing it to view gadgets too vintage, far flung, or faint for the Hubble Space Telescope. This allows investigations all through many fields of astronomy and cosmology, which includes commentary of the primary stars, the formation of the first galaxies, and unique atmospheric characterization of likely liveable exoplanets.

The U.S. National Aeronautics and Space Administration (NASA) led JWST's design and improvement and partnered with number one groups:

the European Space Agency (ESA) and the Canadian Space Agency (CSA). The NASA Goddard Space Flight Center (GSFC) in

Maryland managed telescope improvement, at the identical time due to the fact the Space Telescope Science Institute in Baltimore at the Homewood Campus of Johns Hopkins University presently operates JWST. The primary contractor for the project have turn out to be Northrop Grumman. The telescope is known as after James E. Webb, who became the administrator of NASA from 1961 to 1968 for the duration of the Mercury, Gemini, and Apollo applications.

The James Webb Space Telescope changed into released on 25 December 2021 on an Ariane 5 rocket from Kourou, French Guiana, and arrived at the Sun–Earth L2 Lagrange factor in January 2022. The first JWST photograph grow to be launched to the overall public thru a press conference on 11 July 2022.

JWST's primary mirror includes 18 hexagonal replicate segments crafted from gold-plated beryllium, which combined create a 6.Five-meter- diameter (21 ft) reflect, in comparison with Hubble's 2.Four m (7 feet 10 in). This offers JWST a mild-amassing region of approximately 25 square meters, approximately six instances that of Hubble. Unlike Hubble, which observes in the near ultraviolet and visible (zero.1 to 0.Eight μm), and close to

infrared (0.Eight -- 2.Five μm)[14] spectra, JWST observes a decrease frequency variety, from prolonged-wavelength visible mild (purple) through mid-infrared (0.6–28.Three μm). The telescope wants to be stored as a substitute cold, under 50 K (−223 °C; −370 °F), such that the infrared mild emitted via the telescope itself does no longer interfere with the accumulated mild. It is deployed in a sun orbit close to the Sun–Earth L2 Lagrange element, about 1.Five million kilometers (930,000 mi) from Earth, in which its 5-layer sunshield protects it from warming via the

Sun, Earth, and Moon.

Initial designs for the telescope, then named the Next Generation Space Telescope, started in 1996. Two concept studies were commissioned in 1999, for an potential launch in 2007 and a US$1 billion fee variety. The software program became plagued with huge fee overruns and delays; a number one redecorate in 2005 brought about the cutting-edge approach, with introduction finished in 2016 at a total price of US$10 billion. The immoderate-stakes nature of the release and the telescope's complexity were remarked upon via way of using the media, scientists, and engineers.

Features

The mass of the James Webb Space Telescope is half of of that of the Hubble Space Telescope. The JWST has a 6.Five m (21 toes)-diameter gold-protected beryllium primary duplicate manufactured from 18 separate hexagonal mirrors. The reflect has a complicated place of 26.Three m2 (283 squareToes), of which zero.Nine m2 (9.7 rectangular toes) is obscured thru the secondary assist struts,[15] giving a whole gathering location of 25.Four m2 (273 squareToes).

This is over six times large than the collection region of Hubble's 2.Four m (7.Nine feet) diameter mirror, which has a amassing vicinity of 4.0 m2 (forty 3 rectangular ft). The replicate has a gold coating to offer infrared reflectivity and that is blanketed via a thin layer of glass for sturdiness.

JWST is designed regularly for near-infrared astronomy, but also can see orange and crimson visible moderate, similarly to the mid-infrared location, relying on the device being used. It can come upon gadgets as a extraordinary deal as one hundred times fainter than Hubble can, and devices misplaced in advance within the records of the universe, another time to redshift z≈20 (about 100 80 million years of cosmic time after the Big Bang). For evaluation, the earliest stars are idea to have lengthy-installed among z30 and z≈20 (one hundred–100 eighty million years cosmic time), and the number one galaxies may additionally additionally moreover have prolonged-hooked up spherical redshift $z \approx 15$ (about 270 million years cosmic time). Hubble is not capable of look further lower again than very early reionization at approximately $z \approx eleven.1$ (galaxy GN-z11, 4 hundred million years of cosmic time).

The layout emphasizes the close to to mid-infrared for severa motives:

excessive-redshift (very early and a ways off) items have their visible emissions shifted into the infrared, and therefore their slight can be determined nowadays simplest thru infrared astronomy;

infrared moderate passes more with out problems through dirt clouds than visible moderate;

less warmth gadgets which includes debris disks and planets emit maximum strongly inside the infrared;

These infrared bands are hard to have a look at from the ground or thru way of current location telescopes together with Hubble.

Rough plot of Earth's atmospheric absorption (or opacity) to numerous wavelengths of electromagnetic radiation, which encompass seen lightGround-primarily based definitely certainly telescopes need to glance through Earth's surroundings, this is opaque in many infrared bands (see decide at right). Even in which the surroundings is obvious, a number of the purpose chemical materials, together with water, carbon dioxide, and methane, moreover exist inside the Earth's surroundings, quite complicating evaluation. Existing location telescopes along side Hubble can't test those bands due to the reality their mirrors are insufficiently cool (the Hubble replicate is maintained at approximately 15 °C [288 K; 59 °F]) due to this that the telescope itself radiates strongly within the applicable infrared bands.

JWST also can take a look at devices inside the Solar System at an angle of greater than eighty five° from the Sun and having an apparent angular charge of movement a lot less than zero.03 arc seconds in keeping with second.[a] This includes Mars, Jupiter, Saturn, Uranus, Neptune, Pluto, their satellites, and comets, asteroids and minor planets at or past the orbit of Mars. JWST has the close to-IR and mid-IR sensitivity for you to test clearly all appeared Kuiper Belt Objects. In addition, it can have a take a look at opportunistic and unplanned dreams indoors forty eight hours of a choice to obtain this, which includes supernovae and gamma ray bursts.

Location and orbit

JWST operates in a halo orbit, circling round a problem in place known as the Sun–Earth L2 Lagrange detail, approximately 1,500,000 km (930,000 mi) beyond Earth's orbit during the Sun. Its actual function varies between approximately 250,000 and 832,000 km (a hundred and fifty five,000–517,000 mi) from

L2 as it orbits, preserving it out of every Earth and Moon's shadow. By way of evaluation, Hubble orbits 550 km (340 mi) above Earth's floor, and the Moon is greater or tons a lot much less four hundred,000 km (250,000 mi) from Earth. Objects near this Sun–Earth L2 trouble can orbit the Sun in synchrony with the Earth, permitting the telescope to stay at a more or less constant distance[28] with non-prevent orientation of its sunshield and device bus inside the course of the Sun, Earth and Moon. Combined with its large shadow-heading off orbit, the telescope can simultaneously block incoming warm temperature and slight from all three of these our our our bodies and keep away from even the smallest adjustments of temperature from Earth and Moon shadows that would have an impact on the form, however but keep uninterrupted solar energy and Earth communications on its solar-going through facet. This arrangement keeps the temperature of the spacecraft consistent and underneath the 50 K (−223 °C; −370 °F) critical for faint infrared observations.

Chapter 11: Sunshield Safety

The James Webb Space Telescope (JWST) sunshield is a passive thermal manipulator deployed placed up-release to defend the telescope and instrumentation from the mild and warmth of the Sun, Earth, and Moon. By maintaining the telescope and gadgets in eternal shadow, it allows them to relax to their format temperature of forty kelvins (−233 °C; −388 °F). Its complex deployment changed into successfully finished on January four, 2022, ten days after release, while it emerge as greater than zero.Eight million kilometers (500,000 mi) a ways from Earth.

The JWST sunshield is set 21 m × 14 m (69 ft × forty six feet), extra or plenty much less the dimensions of a tennis court docket, and is truly too massive to be in shape in any modern-day rocket. Therefore, it changed into folded as much as in form inside the fairing of the discharge rocket and function become deployed positioned up-release, unfolding five layers of metallic-included plastic. The first layer is the most essential, and every consecutive layer decreases in length. Each layer is made from a thin (50 micrometers for the first layer, 25 micrometers for the others) Kapton membrane blanketed with aluminum for reflectivity. The outermost Sun-going via layers have a doped-silicon coating which offers it a crimson coloration, toughens the guard, and permits it to copy warmth. The thickness of the aluminum coating is set one hundred nanometers, and the silicon coating is even thinner at approximately 50 nanometers. The sunshield phase consists of the layers and its deployment mechanisms, which furthermore includes the trim flap.

Overview

In this artist's view, a stylized portrayal of the orientation of the telescope, indicates how the sunshield blocks daytime from heating the principle mirror. (not to scale)

To make observations inside the close to and mid infrared spectrum, the JWST want to be saved very cold (beneath 40 K (−233 °C; −388

°F)), otherwise infrared radiation from the telescope itself could weigh down its devices. Therefore, it makes use of a large sunshield to damten mild and warmth from the Sun, Earth, and Moon, and its function in the Sun-Earth L2 Lagrange problem keeps all three our our our bodies at the identical aspect of the spacecraft continually. Its halo orbit round L2 avoids the shadow of the Earth and Moon, maintaining a everyday surroundings for the sunshield and solar arrays.

Infrared is warm temperature radiation. In order to look the faint glow of infrared warmth from a long way flung stars and galaxies, the telescope want to be very cold. If daylight or the first-rate and cushty glow of the Earth heated the telescope, the infrared slight emitted with the aid of using the telescope may possibly outshine its dreams, and it wouldn't be capable of seeing some thing.

The temperature variations a number of the new and bloodless aspects of the James Webb Space Telescope 5-layer sunshield.

The sunshield acts as huge parasol allowing the principle reflect, optics, and gadgets to passively cool to forty kelvins (−233 °C; −388 °F) or cooler, and is one of the permitting era a good way to permit the JWST to characteristic. The kite-traditional sunshield is about 21 via 14 meters (sixty nine through 46 ft) in period,[11] massive enough to coloration the principle replicate and secondary mirror, leaving simplest one device, the MIRI

(Mid-Infrared Instrument), in want of extra cooling.[6] The sunshield acts as a V-groove radiator and reasons a temperature drop of 318 K (318 °C, 604 °F)[12] from the the front to again.[11] In operation the protect gets keep of approximately two hundred kilowatts of solar radiation, however first-class bypass 23 milliwatts to the opportunity aspect.

The sunshield has five layers to mitigate the conduction of warmth. These layers are manufactured from the polyimide film Kapton E, that is strong from −269 to four hundred °C (−450 to 750 °F). However the skinny films are touchy - unintended tears in the end of checking out in 2018 have been a number of the factors delaying the JWST undertaking, and Kapton is thought to degrade after long time publicity to Earth conditions. The solar-dealing layer is .05 mm (0.002 in) thick, and the opportunity layers are .Half of mm (0.001 in) thick. All layers are covered on every component with 100 nm of aluminum, and the Sun-dealing with elements of the outermost layers are also covered with 50 nm of silicon "doped" with different factors.

This enables the cloth live to tell the tale in vicinity, radiate more warmth, and to behavior energy, so a static price does no longer constructing up at the layers.

Test unit of the sunshield stacked and improved at the Northrop Grumman facility in

California, 2014 Each layer has a slightly considered one in every of a type shape and period. Layer 5 is the nearest to the number one mirror and is the smallest. Layer 1 is closest to the Sun and is larger and flatter. The first layer blocks ninety% of the warmth, and every successive layer blocks greater warmth, this is contemplated out the edges. The sunshield allows the optics to stay in shadow for pitch angles of +five° to −45° and roll angles of +five° to −5°. The layers are designed with Thermal Spot Bond (TSB), with a grid pattern bonded to every layer at durations. This permits a rip or hole to stop growing in length.

Design and manufacture

Coupons of Sunshield check fabric being tested to appearance how they carry out, 2012

Northrop Grumman designed the sunshield for NASA. The sunshield is designed to be folded twelve times so it could in shape within the Ariane five rocket's four.Fifty seven m (15.Zero toes) diameter by way of way of sixteen.19 m (fifty 3.1 feet) shroud. When it deployed at the L2 factor, it unfold out to 21.197 m × 14.162 m (69.Fifty four feet × 46.Forty six toes). The sunshield have become hand-assembled at ManTech (NeXolve) in Huntsville, Alabama in advance than it have turn out to be introduced to Northrop Grumman in Redondo Beach, California for attempting out.[18] During launch it changed into wrapped at some stage inside the Optical Telescope Element after which later spread out.[11] The sunshield became consider to be unfold out approximately one week after release. During development the sunshield layer material turns into tested with warm temperature, cold, radiation, and high-pace micro impacts.

Components of the sunshield encompass:

Core

Front and aft four-bar linkage

Aft form assembly

Momentum trim tab (the tab is installation to the aft form meeting)

Aft spreader bars (spreads layers in the rear)

Forward shape meeting

Forward spreader bars

Mid-booms (one on each problem)

Mid-spreader bars (spreads the 5 layers aside)

Two earlier and aft bipod release lock assemblies

The bipod release lock assemblies are in which the sunshield segment linked to the NOTE on the equal time as it modified into folded up in the direction of launch.

There are six spreader bars that increased to separate the layers of the sunshield, which has six components.

Trim flap/momentum trim tab

The sunshield segment furthermore includes a trim flap on the give up of a sunshield deployment boom. This is likewise called the momentum trim tab.

The trim tab lets in stability out of sun strain because of photons placing on the sunshield. If this stress is choppy, the spacecraft will commonly have a tendency to rotate, requiring its response wheels (placed inside the spacecraft bus) to correct and hold JWST's orientation in vicinity. The reaction wheels, in turn, will ultimately end up saturated and require fuel to desaturate, potentially restricting spacecraft lifetime. The trim tab, via assisting preserve the stress balanced and because of this proscribing fuel utilization, extends the jogging existence of the telescope.

Layers

The 5 layers of the JWST sunshield being tested in 2013 The layers are designed so the Sun, Earth, and Moon shine on layer one nearly without a doubt, occasionally a tiny part of layer , and

On the opposite facet that the telescope elements most efficiently see layer 5 and occasionally a tiny amount of layer 4. The separation between layers, within the vacuum of area, prevents warmth switch with the useful resource of conduction and aids in radiating warmth out of the way. Silicon doping of the material motives the red hue.

Deployment

Animation collection for the deployment of the sunshield. To see how the sunshield deployment suits within the complete series of deployments on the spacecraft, see this animation.

The sunshield problem attaches to the primary spacecraft, and its booms boom outward spreading out the warmth shielding and isolating the layers. During launch the protection is folded up; later, while it's far in location, it is cautiously unfurled. When the sunshield is really unfold open, it's miles 14.6 meters (forty eight feet) huge through 21.1 meters (69 ft) lengthy.[20] When the layers are in truth open, they may be opened wider at the edges which lets in them to copy warmness out.

Sunshield deployment shape/devices embody:

telescoping booms

stem deployers

spreader bars

cable drives

There are stem deployers inside the telescoping booms. These are particular electric powered powered cars that, at the equal time as operated, extended the telescopic increase, pulling out the folded sunshield. The telescopic booms are known as the MBA, or mid-increase assemblies. At the give up of every MBA is a spreader bar.

After a achievement launch on 2021 December 25 from the Guiana Space Center, the positioned up-release deployment of the JWST sunshield proceeded as follows.

On December 31, 2021, the floor corporation on the Space Telescope Science Institute in Baltimore, Maryland started the deployment of the 2 telescoping "mid-booms" from the left and right factors of the observatory, pulling the five sunshield membranes out of their folded garage within the fore and aft pallets, that have been decreased 3 days in advance. Deployment of the left element boom (close to pointing

path of the primary reflect) end up not on time even as undertaking manage did not to start with get hold of confirmation that the sunshield cowl had absolutely rolled up. After looking at greater information for affirmation, the team proceeded to growth the booms. The left thing deployed in three hours and 19 mins; the proper issue took 3 hours and 40 minutes.With that step, Webb's sunshield resembled its complete, kite-authentic form and prolonged to its complete forty seven-foot width. Commands to break up and tension the membranes have been to check.

After taking New Year's Day off, the ground employer delayed sunshield tensioning with the useful useful resource of sooner or later to allow time to optimize the power output of the observatory's array of solar panels and to modify the orientation of the observatory to sit back the marginally-hotter-than-anticipated sunshield deployment cars. Tension of layer one, closest to the Sun and largest of the five within the sunshield, began on 2022 January three and changed into finished at three:forty eight p.M. EST. Tensioning of the second and 1/three layers started out at four:09 p.M. EST and took 2 hours and 25 mins. On January four, 2022, controllers efficaciously tensioned the ultimate layers, four and five, completing the challenge of deploying the JWST sunshield at eleven:fifty nine a.M. EST.

Chapter 12: Optical Telescope Element

Optical Telescope Element (OTE) is a sub-phase of the James Webb Space Telescope, a large infrared location telescope released on 25 December 2021, which incorporates its primary mirror, secondary mirrors, the framework and controls to assist the mirrors, and numerous thermal and other systems.

The OTE collects the slight and sends it to the technological understanding devices in Webb's Integrated Science Instrument Module.The VOTE has been in evaluation to being the "eye" of the telescope and the backplane of it to being the "spine".

The number one mirror is a tiled assembly of 18 hexagonal factors, every 1.32 meters (4.Three toes) from flat to flat. This aggregate yields an effective aperture of 6.Five meters (21 ft) and a entire amassing floor of 27 rectangular meters (290 squareToes). Secondary mirrors entire anastigmatic imaging optics with effective f/20 focal ratio and focal length of 131.Four meters (431 ft),

The fundamental three-reflect telescope is a Korsch-type layout, and it feeds into the Aft Optics Subsystem (part of OTE), which in flip feeds into the Integrated Science Instrument Module which holds the technological information gadgets and pleasant guidance sensor.

The specific principal sections of the JWST are the Integrated Science Instrument Module (ISIM) and the Spacecraft Element (SE), which includes the spacecraft bus and sunshield.[6] The additives of OTE have been included thru L3 Harris Technologies to form the final gadget.

Overview

The OTE combines a massive amount of the optics and structural additives of the James Webb Space Telescope, together with the Main mirror. It moreover has the fine steering mirror, which offers that final particular pointing, and it in fact works on the side of the superb steerage sensor and wonderful controls structures and sensors within the spacecraft bus.

The most important mirror segments are aligned extra or plenty much less via the use of a tough phasing set of guidelines. Then for finer alignment, unique optical devices inner NIRCam are used to act a section retrieval approach, to accumulate designed wavefront mistakes of an entire lot plenty a good deal less than one hundred fifty nm. To characteristic as a focusing reflect successfully the 18 important mirror segments need to be aligned very cautiously to perform as one. This wants to be completed in outer regions, so significant attempting out on Earth is needed to make sure that it'll artwork well. To align every reflect segment, it's miles set up to 6 actuators that would adjust that section in five nm steps. One cause the replicate have become divided into segments is that it cuts down on weight, because of the fact a mirror weight is related to its duration, which is also one of the reasons beryllium grow to be decided on as the mirror fabric because of its low weight. Although within the basically weightless environment of area the reflector will weigh hardly ever some factor, it wishes to be very stiff to preserve its shape. The Wavefront sensing and manipulating sub-device is designed to make the 18 phase primary replicate behave as a monolithic (single-piece) mirror, and it does this in detail by way of the usage of using actively sensing and correcting for mistakes. There are nine distance alignment techniques that the telescope goes through to advantage this. Another essential trouble to the modifications is that the primary mirror backplane assembly is normal. The backplane meeting is product of graphite composite, invar, and titanium.

The ADIR, Aft Deployable Infrared Radiator is a radiator in the back of the principle reflect, that permits maintain the telescope cool.There are ADIR's and they may be products of immoderate-purity aluminum. There is a very unique black coating at the radiators that allows them to emit warm temperature into the location.

Testing of the Aft Optic Subsystem in 2011, which incorporates the Tertiary (0.33) mirror and Fine Steering Mirror

Some primary additives of the OTE regular with NASA:

Primary reflect (18 segments)

Secondary mirror (seventy four cm (29 in) diameter)

Tertiary replicate (0.33) (in Aft Optics Subsystem)

Fine Steering Mirror (in Aft Optics Subsystem)

Telescope shape

number one mirror backplane assembly

important backplane manual fixture (BSF)

secondary reflect assist shape

deployable tower array

Thermal Management Subsystem

Aft Deployable ISIM Radiator (ADIR)

Wavefront sensing and manage

The Aft Optics Subsystem includes the Tertiary mirror and the Fine Steering Mirror.

The metal beryllium modified into decided on for some of motives which incorporates weight, but also for its low-temperature coefficient of thermal increase in evaluation to glass. Furthermore beryllium is not magnetic and a superb conductor of energy and warmth. Other infrared telescopes that have used beryllium mirrors encompass IRAS, COBE, and Spitzer. The Subscale Beryllium Model Demonstrator (SBMD) come to be successfully examined at cryogenic temperatures, and one of the troubles changed into floor roughness at low kelvin numbers. The beryllium mirrors are coated with a completely tremendous layer of gold to mirror infrared mild. There are 18 hexagonal segments which is probably grouped collectively to create a single replicate with an sizeable diameter of 6.Five meters

DTA

The Deployable Tower Assembly (DTA) is wherein OTE connects with the relaxation of the telescope which incorporates the spacecraft bus. During stowage there may be every exclusive attachment issue for the folded sunshield higher up at the GATE At the bottom of the OTE is the essential Deployable Tower Assembly (DTA).Aspect which connects the VOTE to the spacecraft bus. It has to increase to allow the Sunshield (JWST) to spread out, permitting the distance most of the five layers to increase. The sunshield phase has a couple of additives, along side six spreaders on the periphery to unfold the layers out at the six extremities.

During release the DTA is shrunk down, however need to growth at the proper second.The prolonged DTA form permits the sun defend layers to be definitely unfold-out.[15] The DTA need to moreover thermally isolate the bloodless section of the OTE from the latest spacecraft bus.The Sunshield will shield the OTE from direct daytime and decrease the thermal radiation hitting it, however each different detail is the OTE's physical connection to the rest of the spacecraft. (see Thermal conduction and Heat switch) Whereas the sunshield stops the telescope getting warm because of radiated warmth from the Sun, the DTA need to insulate the telescope from the warm temperature of the relaxation of the shape, just like the manner an insulated pan deal with protects from the warm temperature of a selection.

The DTA extends via way of using telescoping tubes which could slide amongst each specific on rollers. There is an internal tube and an outer tube. The DTA is extended with the useful useful aid of an electric powered motor that rotates a ball screw nut which pushes the two tubes aside. When the DTA is certainly deployed it's miles three meters (10 ft) extended.[16] The DTA tubes are a fabricated from graphite-composite carbon fiber, and it is supposed that they may be able to persevering with to exist in vicinity.

A one 6th scale test version of the number one reflect

Achieving a on foot vital reflect have emerge as one of the pleasant demanding situations of JWST development.Part of the JWST development protected validating and finding out JWST on severa testbeds of various capabilities and sizes.

Some sorts of development devices encompass pathfinders, check beds, and engineering take a look at gadgets. Sometimes an single item can be used for tremendous competencies, or it may not be a physical created object in any respect, but as an opportunity to a software program program software software simulation. The NEXUS region telescope became an entire location telescope, basically a scaled down JWST but with some changes collectively with simplest 3 replicate segments with one folding out for a immoderate replicate diameter of .Eight meters (9.2 feet). It have become lighter, so it modified into expected it is able to be launched as early as 2004 on a Delta 2 release rocket.[28] The format have become canceled at the surrender of 2000. At that thing NGST/JWST have become regardless of the reality that a eight-meter (26 toes) design, with a place of 50 m2 (540 rectangular feet), a few years later this changed into reduced sooner or later to the 6.Five-meter (21 toes) layout, with an area of 25 m2 (270 square toes).

OTE Pathfinder

One a part of JWST improvement become the manufacturing of the Optical Telescope Element Pathfinder. The OTE pathfinder makes use of more reflect segments, and similarly secondary reflect, and places together severa systems to permit finding out of severa elements of the phase, along with Ground Support Equipment. This lets in the GSE being used at the JWST itself in a while, and allows trying out of replicate integration. OTE pathfinder has 12 in area of 18 cells compared to the overall telescope, however it does encompass a take a look at of the backplane shape.

Additional checks/models

There are many check articles and developmental demonstrators for the introduction of JWST. Some critical ones had been early demonstrators that confirmed that hundreds of crucial generation of JWST had been feasible. Others test articles which might be vital for danger mitigation, essentially lowering the general hazard of this device thru practicing on a few factor other than the real flight spacecraft.

Another testbed, the Test Bed Telescope, have emerge as a 1/sixth scale version of the primary mirror, with polished segments and on foot actuators, operating at room temperature, and used to test all the techniques for aligning the segments of JWST. Another optics testbed is known as JOST, which stands for JWST Optical Simulation Testbed, and makes use of an MEMS with hexagonal segments to simulate the levels of freedom of the number one reflect alignment and phasing.

The Subscale Beryllium Model Demonstrator (SBMD) became fabricated and examined through 2001 and set up permitting generation for what have become quickly Christened the James Webb Space Telescope, formerly the Next Generation Space Telescope (NGST).[18] The SBMD have emerge as a 1/2 of of-meter diameter replicate crafted from powdered beryllium. The weight of the reflect is then reduced thru a replicate-making device called "slight-weighting", in which material is eliminated with out disrupting its reflecting capacity, and in this example 90% of the SBMD mass turns into eliminated. It have become then installation to a inflexible backplane with titanium bipod flexures and underwent numerous checks. This protected freezing it all the manner right down to the low temperatures required and seeing the way it behaved optically and physically. The assessments were done with the Optical

Testing System (aka the OTS) which have become created specifically to check the SBMD. The SBMD needed to meet the necessities for a location-based totally completely reflect, and those commands have been essential to the development of the JWST. The checks were accomplished on the X-Ray Calibration Facility (XRCF) at Marshall Space Flight Center (MSFC) within the U.S. State of Alabama.

The Optical Testing System (OTS) needed to be advanced to check the SBMD (the NGST replicate prototype) under cryogenic vacuum conditions. The OTS included a WaveScope Shack-Hartmann sensor and a Leica Disto Pro distance duration tool.

Some JWST generation Testbeds, Pathfinders, and so forth.

Chapter 13: Scientific Devices

The Integrated Science Instrument Module (ISIM) is a framework that gives electric powered electricity, computing resources, cooling functionality in addition to structural balance to the Webb telescope. It is made with a bonded graphite-epoxy composite related to the underside of Webb's telescope form. The ISIM holds the 4 technology devices and a guide digicam.

NIRCam (Near Infrared Camera) is an infrared imager which has spectral insurance starting from the brink of the scene (0.6 μm) thru to the near infrared (5 μm). There are 10 sensors each four megapixels. NIRCam serves due to the fact the observatory's wavefront sensor, which is required for wavefront sensing and manipulating sports activities sports, used to align and hobby the number one reflect segments. NIRCam grow to be constructed by using the usage of a group led via the University of Arizona, with fundamental investigator Marcia J. Rieke.

NIRSpec (Near Infrared Spectrograph) performs spectroscopy over the equal wavelength range. It became constructed through the European Space Agency at ESTEC in Noordwijk, Netherlands. The main improvement corporation includes individuals from Airbus Defence and Space, Ottobrunn and Friedrichshafen, Germany, and the Goddard Space Flight Center; with Pierre Ferruit (École normale supérieure de Lyon) as NIRSpec project scientist. The NIRSpec layout presents three searching modes: a low-choice mode the usage of a prism, an R~1000 multi-item mode, and an R~2700 essential area unit or lengthy-slit spectroscopy mode. Switching of the modes is completed through operating a wavelength preselection mechanism referred to as the Filter Wheel Assembly, and choosing a corresponding dispersive detail (prism or grating) using the Grating Wheel Assembly mechanism. Both mechanisms are based on the a fulfillment ISOPHOT wheel mechanisms of the Infrared Space Observatory. The multi-item mode is primarily based mostly on a complex micro-shutter mechanism to permit for simultaneous observations of loads of individual gadgets anywhere in NIRSpec's location of view. There are sensors, every of four megapixels.

MIRI (Mid-Infrared Instrument) measures the mid-to-prolonged-infrared wavelength variety from 5 to 27 μm. It includes every a mid-infrared digital camera and an imaging spectrometer.[49] MIRI was developed as a collaboration amongst NASA and a consortium of European global locations, and is led via manner of George Rieke (University of Arizona) and Gillian Wright (UK Astronomy Technology Centre, Edinburgh, Scotland). The temperature of the MIRI has to not exceed 6 K (−267 °C; −449 °F): a helium fuel mechanical cooler sited at the fine and comfortable side of the environmental defend offers this cooling.

FGS/NIRISS (Fine Guidance Sensor and Near Infrared Imager and Slitless Spectrograph), led with the useful aid of the Canadian Space Agency underneath assignment scientist John

Hutchings (Herzberg Astronomy and Astrophysics Research Centre), is used to stabilize the road-of-sight of the observatory in the path of science observations. Measurements with the aid of manner of manner of the FGS are used each to control the overall orientation of the spacecraft and to pressure the extremely good guidance reflect for photo stabilization. The Canadian Space Agency additionally provided a Near Infrared Imager and Slitless Spectrograph (NIRISS) module for astronomical imaging and spectroscopy in the 0.Eight to 5 μm wavelength range, led through way of number one investigator René Doyon at the Université de Montréal. Although they'll be frequently referred together as a unit, the NIRISS and FGS serve simply one-of-a-type capabilities, with one being a systematic device and the opposite being part of the observatory's manual infrastructure.

Chapter 14: Spacecraft Bus

Technician's artwork on a mock-up of the JWST spacecraft bus in 2014 The spacecraft bus is the primary assist element of the James Webb Space Telescope, released on 25 December 2021. It hosts a mess of computing, verbal exchange, propulsion, and structural components. The 3 unique factors of the JWST are the Optical Telescope Element (OTE), the Integrated Science Instrument Module (ISIM) and the sunshield. Region three of ISIM is also in the spacecraft bus. Region three includes the ISIM Command and Data Handling subsystem and the Mid-Infrared Instrument (MIRI) cryocooler.

The spacecraft bus has to structurally assist the 6.Five ton space telescope, on the same time as weighing 350 kg (770 lb). It is made extra regularly than not of graphite composite fabric. It become assembled in the U.S. State of California thru 2015, after which it had to be included with the rest of the space telescope crucial as lots as its deliberate 2018 release. The bus can offer pointing precision of 1 arcsecond (1/3600°) and isolates vibration down to 2 milliarcseconds. The nice pointing is finished via the JWST high-quality steerage reflect, obviating the want to bodily circulate the whole replicate or bus.

The spacecraft bus is at the Sun-going via "warmness" issue and operates at a temperature of approximately 3 hundred kelvins (eighty °F, 27 °C). Everything at the Sun-dealing component need to be able to dealing with the thermal situations of JWST's halo orbit, which has one element of non-save you daylight and the alternative shaded thru the spacecraft sunshield.

Another vital element of the spacecraft bus is the massive computing, memory garage, and communications device. The processor and software utility direct statistics to and from the gadgets, to the solid-u . S . Memory center, and to the radio system which can supply facts once more to Earth and accumulate commands. The laptop moreover controls the pointing and movement of the spacecraft, taking in sensor statistics from the gyroscopes and movie star tracker, and sending the crucial instructions to the response wheels or thrusters.

Overview

In the Diagram of the spacecraft bus. The solar panel is in inexperienced and the mild purple residences are radiator sunglasses.

The bus is a carbon fiber field that houses a massive form of most important systems that preserve the telescope functioning, on the side of the solar panels and computer systems. It additionally consists of the MIRI cooler and a few ISIM electronics.

There are six number one subsystems in the spacecraft bus:

Electrical Power Subsystem

Attitude Control Subsystem

Communication Subsystem

Command and Data Handling Subsystem (C&DH)

Command Telemetry Processor

Solid State Recorder (SSR)

Propulsion Subsystem

Thermal Control Subsystem

The spacecraft bus has big name trackers, six response wheels, and the propulsion systems (gasoline tank and thrusters). Two most essential obligations are pointing the telescope and retaining station preservation for its metastable L2 halo orbit.

Command and Data Handling (C&DH)

The Command and Data Handling device consists of a pc, the Command Telemetry Processor (CTP), and a facts garage unit, the Solid State Recorder (SSR),[2] with an ability of 50 eight.Nine GB.

Communications

The communications dish that would factor at Earth is established to the bus.: Fig 1 There is Ka-band and S-band radio communication. The Common Command and Telemetry System is primarily based totally on the Raytheon ECLIPSE system. The gadget is designed to talk with NASA's Deep Space Communication Network. The maximum essential Science and Operations Center is the Space Telescope Science Institute within the U.S. Nation of Maryland.

Rocket engines, mind-set manipulate

The JWST makes use of sorts of thrusters. The Secondary Combustion Augmented Thrusters (SCAT) use hydrazine (N_2H_4) and the oxidizer dinitrogen tetroxide (N_2O_4) as propellants. There are 4 SCATs in pairs. One pair is used to propel the JWST into orbit, and the opportunity plays station retaining in orbit. There are also 8 Monopropellant Rocket Engines (MRE-1), so known as because of the fact they use the only hydrazine as gas. They are used for thoughts-set manipulate and momentum unloading of the reaction wheels.

JWST has six reaction wheels for mind-set manage, spinning wheels that allow the orientation to be modified with out the usage of propellant to exchange momentum.

Finally, there are titanium helium tanks to offer unregulated pressurant for all propellants.[citation needed]

To hit upon adjustments in course JWST makes use of hemispherical resonator gyroscopes (HRG). HRGs are expected to be extra reliable than the gas-bearing gyroscopes that have been a reliability problem on the Hubble Space Telescope. They can not detail as finely, but, that is overcome with the beneficial resource of the JWST first-class steerage replicate.

Thermal

Thermal structures on the bus encompass the Deployable Radiator Shade Assemblies. There are , one vertical (DRSA-V) and one horizontal

(DRSA-H), for vertical and horizontal respectively (with admire to the coordinate device of the spacecraft bus). The membrane that makes up the DRSA is a blanketed Kapton membrane. Other thermal factors at the outdoor encompass a small radiator for the battery. There is likewise a narrow lower-consistent radiator color, moreover a manufactured from covered Kapton membrane. The coating of the membrane is silicon and VPA. Other areas of the out of doors are protected with JWST multi-layer insulation (MLI).

Electrical Power Subsystem (EPS)

The Electrical Power Subsystem offers power to the JWST spacecraft.[19] It includes a hard and rapid of solar panels and rechargeable batteries,[19][20] a sun array regulator (SAR), a power manipulate unit (PCU), and a telemetry acquisition unit (TAU).

The sun panels convert sunlight hours proper away into power.[19] This uncooked energy is fed to the SAR which includes four redundant dollar converters every working with a most-strength element tracking (MPPT) set of regulations. While the output voltage isn't tightly regulated, the dollar converters will now not permit the spacecraft's maximum vital bus voltage to drop under about 22 volts, or upward push above approximately 35 volts. With every technological facts tool and all useful resource circuits "on" simultaneously, approximately 3 of the four redundant converters need to manipulate all of the energy required. Typically one or converters want to be walking at a time with the opposite on energetic standby.

The Power Control Unit (PCU) consists specifically of digital switches that flip every technological expertise device or useful resource tool on or off below the supervisor of the essential laptop. Each switch lets in power to go with the flow to its decided on device from the SAR. Communication with the critical laptop is thru a 1553 bus. In addition to the power switches, processors for the SAR MPPT set of rules are placed inside the PCU, on the factor of some telemetry processors, processors to find out while the spacecraft has disconnected from the release higher diploma, and some cryo-cooler controllers.

The Telemetry Acquisition Unit (TAU) consists of virtual switches for severa heaters for the "warmness" factors of the telescope. In addition, there are switches for the deployment actuators, and the majority of the telemetry processors (e.G. Measuring temperatures, electric powered powered energy, fuel ranges, etc.). The TAU communicates with the precious pc through 1553 bus.

Both the PCU and TAU encompass clearly redundant systems with one energetic at the same time as the possibility is in standby mode or off, completely. The rechargeable batteries of JWST are the lithium-ion kind.[20] The batteries use the Sony 18650 hard carbon mobile generation. The batteries are designed to undergo spaceflight, and need to preserve 18,000 charge-discharge cycles. Each sun panel form help is honey-comb carbon fiber composite.[citation needed]

Some early configurations of the bus had solar panel wings, one on each aspect. Part of the JWST software format modified into permit wonderful layout variations to "compete" with every one-of-a-kind.

Structure

Although the bus will carry out within the weightless environment of the outer region, throughout launch it want to live on at the same of forty five masses.The form can aid sixty four instances its private weight.

The spacecraft shape gives us of a of the art work abilities to aid the James Webb Space Telescope's first light assignment.

A Webb Telescope spacecraft supervisor as quoted through Composites World

The spacecraft bus is installed to the Optical Telescope Element and sunshield via the Deployable Tower Assembly.The interface to the discharge automobile in on out of doors; taking the form of a cone, it collectively with the payload adapter transmits the weight and acceleration forces outward release vehicle partitions.

The shape of the bus partitions are manufactured from carbon fiber composite and graphite composite.

The bus is 3,508 mm (eleven.509 toes) lengthy with out the sun arrays. From one fringe of a extended radiator colour to every different it's far 6,775 mm (22.228 ft); this consists of the period of the 2 -meter-massive radiator solar shades. The tail-dragger sun array is five,900 mm (19.Four toes) however it's far normally at an altitude of 20° within the route of the sunshield. The array is inside the the front of the sunshield segments protect deployment increase, which at the forestall of it furthermore has a trim tab linked.

The bus form itself weighs 350 kg (770 lb)

Once JWST is released, it starts offevolved to spread and enlarge to its operating configuration. The plan is that during its first week the deployable tower will boom, as a manner to separate the bus from the top spacecraft by means of approximately 2 meters.

Testing

A software program program software simulation of the Solid-State Recorder advanced for sorting out skills, which supports the overall software program program application simulation of JWST. This is called the JWST Integrated Simulation and Test (JIST) Solid State Recorder (SSR) Simulator, and become used to check flight software program software with SpaceWire and MIL-STD-1553 verbal exchange, as it relates to the SSR.

An Excalibur 1002 Single Board Computer ran a observe a software program program software. The SSR takes a take a look at software program an extension of the JIST software program that's known as JWST Integrated Simulation and Test center (JIST). JUST brings collectively software program software application application simulations of JWST hardware with real JWST software program packages, to permit digital trying out. The simulated SSR changed into created to useful resource in growing a software take a look at version of the JWST, to assist validate and take a look at the flight software program software for the telescope. In precise terms, in place of the use of a actual take a look at hardware version of the SSR, there may be a software program application software that simulates how the SSR works, which runs on some other piece of hardware.

Construction

The Deployable Tower Assembly (DTA) is wherein the spacecraft bus connects to the Optical Telescope Element. When it extends, it moves the bus farther a protracted manner from the principle reflect, developing a place for the sunshield layers.

The spacecraft detail is made through Northrop Grumman Aerospace Systems. The sunshield and Bus are intended to be included in 2017.

In 2014, Northrop Grumman started creation of several spacecraft bus additives consisting of the gyroscopes, gas tanks, and sun panels. On May 25, 2016, the spacecraft's panel integration grow to be finished.[30] The brand new spacecraft bus shape turn out to be completed through October 2015. The spacecraft bus have turn out to be assembled at facilities in Redondo Beach, California inside the United States. The completed spacecraft bus modified into powered on for the number one time in early 2016.

The solar arrays finished a initial format audit in 2012, shifting to the specific layout section. Fuel and oxidizer tanks were shipped out to meeting in September 2015.

In 2015, the communications subsystems, big name trackers, reaction wheels, superb solar sensors, deployment electronics Unit, command telemetry processors, and cord harnesses were brought for introduction.

From 2016 to 2018, there are installations and checks for the telescope and the telescope plus the gadgets, accompanied thru shipping to NASA's Johnson Space Center in Houston, Texas in which give up-to-prevent optical attempting out in a simulated cryo-temperature and vacuum region environment will arise... Then all of the elements can be shipped to Northrop Grumman for the very last assembly and test out, then to French Guiana for launch.

For release, the spacecraft bus is installed to the Ariane 5 on a Cone 3936 plus ACU 2624 decrease cylinder and clamp-band. It is a contained launch fairing, four.Fifty seven meters (15 feet) and 16.19 meters (fifty three.1 feet) of usable interior length.

Gyroscopes

There are most critical conventional uses for gyroscopes in a spacecraft: to come across modifications in orientation, and to honestly exchange the orientation.

JWST makes use of a form of gyroscope called a hemispherical resonator gyroscope (HRG). This format has no bearings, rubbing factors, or flexible connections. This isn't always a traditional mechanical gyroscope; as an alternative, an HRG has a quartz hemisphere that vibrates at its resonant frequency in a vacuum. Electrodes locate modifications if the spacecraft movements to accumulate the popular facts on orientation.

The format is expected to have a mean time in advance than failure of 10 million hours. Gyroscopes failed on numerous sports activities at the Hubble Space Telescope and needed to get replaced numerous times. However, the ones had been an distinct layout referred to as a gasoline-bearing gyroscope, which have certain advantages however professional some extended-term reliability problems. JWST may also additionally have six gyroscopes, but fantastic is needed for pointing. JWST does not want specific pointing as it has a Fine Steering Mirror that allows counter small motions of the telescope.

The JWST telescope furthermore has spinning reaction wheels, which may be adjusted to aspect the telescope with out the use of propellant, further to a difficult and fast of small thrusters that might bodily trade the mind-set of the telescope.

The HRG are sensors that provide facts, on the identical time because the response wheels and thrusters are gadgets that bodily trade the orientation of the spacecraft. Together they painted to preserve the telescope within the right orbit and pointed in the favored course.

Integration

The spacecraft bus is integrated into the entire JWST in a few unspecified time within the destiny of producing. The spacecraft bus and the Sunshield segment are combined into what's known as the Spacecraft Element, it really is in turn blended with a mixed form of the Optical Telescope Element and Integrated Science Instrument Module referred to as OTIS.[39] That is the entire observatory, this is mounted to a cone which connects the JWST to the final level of the Ariane five rocket. The spacecraft bus is wherein that cone connects to the relaxation of JWST.

Servicing

JWST isn't supposed to be serviced in region. A crewed venture to repair or beautify the observatory, as become finished for Hubble, might also not currently be feasible,[59] and constant with NASA Associate Administrator Thomas Zurbuchen, however nice efforts, an uncrewed a long way flung mission turned into located to be past current-day generation at the time JWST emerge as designed.

During the extended JWST checking out period, NASA officials referred to the idea of a servicing challenge, however no plans had been brought. Since the a success launch,

NASA has stated that but confined lodges turn out to be made to facilitate destiny servicing missions. These lodges covered unique guidance markers within the shape of crosses on the surface of JWST, to be utilized by some distance flung servicing missions, in addition to refillable gasoline tanks, detachable warmth protectors, and accessible attachment factors.

Software

Ilana Dashevsky and Vicki Balzano write that JWST uses a changed model of JavaScript, known as Nombas ScriptEase 5.00e, for its operations; it follows the ECMAScript famous and "permits for a modular format go along with the flow, wherein on-board scripts name lower-diploma scripts which can be defined as skills".

"The JWST era operations may be driven via ASCII (in place of binary command blocks) on-board scripts, written in a customized version of JavaScript. The script interpreter is administered through manner of manner of the flight software program software, this is written in C++. The flight software software operates the spacecraft and the era gadgets."

Comparison with particular telescopes

Comparison with the Hubble Space Telescope primary reflect

Primary reflect size evaluation among JWST and Hubble

The choice for a large infrared region telescope traces again a few years. In america, the Space Infrared Telescope Facility (later called the Spitzer Space Telescope) have grow to be deliberate on the identical time because the Space Shuttle have emerge as in improvement, and the capability for infrared astronomy modified into states at that factor. Unlike ground telescopes, area observatories have been free from atmospheric absorption of infrared slight. Space observatories spread out an entire "new sky" for astronomers.

S. G. McCarthy and G. W. Audio, 1978.

However, infrared telescopes have a drawback: they need to live noticeably bloodless, and the longer the wavelength of infrared, the colder they want to be. If now not, the historical beyond warmth of the device itself overwhelms the detectors, making it effectively blind. This may be conquer via using cautious spacecraft format, mainly with the aid of using placing the telescope in a dewar with a truely cold substance, together with liquid helium. The coolant will slowly vaporize, proscribing the life of the tool from as quick as some months to a few years at maximum.

In some instances, it is possible to keep a temperature low enough through the format of the spacecraft to permit near-infrared observations without a deliver of coolant, together with the extended missions of Spitzer Space Telescope and Wide-trouble Infrared Survey Explorer, which operated at decreased capability after coolant depletion. Another instance is Hubble's Near Infrared Camera and Multi-Object Spectrometer

(NICMOS) tool, which began out out the use of a block of nitrogen ice that depleted after multiple years, but end up then replaced at some degree in the STS-109 servicing project with a cryocooler that worked continuously. The James Webb Space Telescope is designed to sit down down returned itself without a dewar, the use of an combination of sun shields and radiators, with the mid-infrared device the usage of a further cryocooler.

Chapter 15: Development and Statistics

Discussions of a Hubble examine-on began out in the Nineteen Eighties, but excessive making plans commenced out out in the early 1990s. The Hi-Z telescope concept become evolved between 1989 and 1994: a totally baffled[b] four m (thirteen toes) aperture infrared telescope that might recede to an orbit at three Astronomical unit (AU).[76] This far flung orbit could have benefited from decreased moderate noise from zodiacal dust. Other early plans referred to as for a NEXUS precursor telescope assignment.

Correcting the incorrect optics of the Hubble Space Telescope in its first years performed a large position inside the beginning of the JWST.[citation needed] In 1993, NASA finished STS-61, the Space Shuttle venture that modified HST's digital digicam and a mounted a retrofit for its imaging spectrograph to atone for the round aberration in its number one reflect.

The HST & Beyond Committee changed into fashioned in 1994 "to test feasible missions and applications for optical-ultraviolet astronomy in area for the primary a few years of the 21st century." Emboldened thru HST's success, its 1996 report explored the idea of a larger and loads less warm, infrared-touch telescope that would be lower back in cosmic time to the begin of the primary galaxies. This excessive-priority technological statistics cause become beyond the HST's functionality because, as a warmness telescope, it is blinded by using manner of the use of infrared emission from its very very very own optical machine.

In addition to pointers to increase the HST undertaking to 2005 and to develop technology for finding planets spherical distinct stars, NASA embraced the leader advice of HST & Beyond for a large, bloodless area telescope (radiatively cooled an extended manner below 0 °C), and started out the making plans way for the future JWST.

Preparation for the 2000 Astronomy and Astrophysics Decadal Survey (a literature assessment produced with the useful resource of america National Research Council that includes identifying studies priorities and making hints for the imminent decade) covered similarly improvement of the medical software program for what became referred to as the Next Generation Space Telescope,and enhancements in applicable generation through NASA. As it matured, reading the shipping of galaxies within the greater youthful universe, and seeking out planets around unique stars – the top goals coalesced as "Origins" through HST & Beyond have become brilliant.

Early development and replanning (2003–2007)

Early whole-scale model on show at NASA Goddard Space Flight Center (2005)

Development become controlled via NASA's Goddard Space Flight Center in Greenbelt, Maryland, with John C. Mather as its venture scientist. The primary contractor come to be Northrop Grumman Aerospace Systems, answerable for growing and building the spacecraft element, which protected the satellite tv for pc bus, sunshield, Deployable Tower Assembly (DTA) which connects the Optical Telescope Element to the spacecraft bus, and the Mid Boom Assembly (MBA) which allows to install the large sunshields on orbit, on the identical time as Ball Aerospace & Technologies has been subcontracted to increase and bring collectively the VOTE itself, and the Integrated Science Instrument Module (ISIM).

Cost increase found out in spring 2005 caused an August 2005 re-making plans. The number one technical consequences of the re-making plans had been massive changes inside the integration and test plans, a 22-month release put off (from 2011 to 2013), and removal of device-degree sorting out for observatory modes at wavelength shorter than 1.7 μm. Other essential skills of the observatory had been unchanged. Following the re-making plans, the venture became independently reviewed in April 2006.

Chapter 16: The James Webb Space Telescope

The James Webb Space Telescope (JWST) is a mainly predicted subsequent-generation vicinity telescope this is expected to redefine our expertise of the universe. It is a joint task of NASA, the European Space Agency (ESA), and the Canadian Space Agency (CSA). The JWST is called after James E. Webb, who served as the second one administrator of NASA from 1961 to 1968, and accomplished a key position within the Apollo software that landed humans at the moon.

The JWST is scheduled to release in 2021 from the Guiana Space Centre in French Guiana, South America, aboard an Ariane 5 rocket. Once in orbit, the JWST can be located at a distance of about 1.Five million kilometers from Earth, at a factor referred to as the second Lagrange point (L2). This vicinity is ideal for astronomical observations because it offers an unobstructed view of the sky, and the telescope is probably protected against the warm temperature and mild emitted thru the Earth, Moon, and Sun.

The JWST is designed to examine the universe within the infrared location of the electromagnetic spectrum. Infrared moderate has longer wavelengths than seen moderate, and it may penetrate via the dust and fuel that regularly obscures astronomical gadgets, permitting us to appearance deeper into location and in addition lower decrease returned in time. The JWST might be able to check the earliest galaxies that authentic after the Big Bang, and study the atmospheres of exoplanets searching out signs and symptoms and symptoms of lifestyles.

The technical abilities of the JWST are fantastic. It has a primary mirror this is 6.Five meters in diameter, this is extra than times the dimensions of the Hubble Space Telescope's number one mirror. The reflect is manufactured from 18 hexagonal segments that may be adjusted for my part to correct for distortions because of modifications in temperature or gravitational forces. The telescope additionally has four clinical contraptions: the Near Infrared Camera (NIRCam), the Near Infrared Spectrograph (NIRSpec), the Mid-Infrared Instrument (MIRI), and the Fine Guidance Sensor/Near InfraRed Imager and Slitless Spectrograph (FGS/NIRISS). Each tool is designed to have a have a look at unique components of the universe, from the formation of stars and galaxies to the look for exoplanets.

One of the maximum thrilling generation goals of the JWST is the have a look at of exoplanets. Exoplanets are planets that orbit stars outdoor of our solar tool, and that they have come to be a number one place of studies in contemporary years. The JWST is probably capable of look at the atmospheres of exoplanets in super element, searching out the chemical signatures of water, methane, and unique molecules that might be indicative of existence. This ought to purpose the discovery of in all likelihood liveable worlds past our sun system.

The JWST also can be able to have a look at the earliest galaxies that formed after the Big Bang. Because the mild from those galaxies has been journeying for billions of years, it has been stretched out to longer wavelengths, making it seen in the infrared vicinity of the spectrum. By analyzing the ones galaxies, astronomers desire to have a take a look at extra approximately how the universe superior and the manner the primary stars and galaxies usual.

The Technical Capabilities of JWST: Instruments and Observations

The James Webb Space Telescope (JWST) is a especially modern-day observatory this is set to push the boundaries of what we are able to take a look at the universe. Its technical competencies are 2d to none, with four clinical gadgets designed to have a look at extraordinary factors of the cosmos. The devices onboard the JWST, and the manner they will permit groundbreaking observations of our universe may be analyzed on this bankruptcy.

The Near Infrared Camera (NIRCam) is one of the number one gadgets onboard the JWST. It is designed to have a observe infrared light from remote galaxies and stars, and to examine the formation of planets and planetary systems. NIRCam is a sensitive device that may discover faint gadgets, and it has a massive place of view, because of this it could test huge regions of the sky right now. It additionally has the capability to take photographs of a couple of gadgets simultaneously, that is essential for reading galaxy clusters and other massive structures within the universe.

The Near Infrared Spectrograph (NIRSpec) is some other key device onboard the JWST. It is designed to have a examine the spectra of remote galaxies and stars, that would inform us loads about their chemical composition, temperature, and movement. NIRSpec has the capability to take a look at as an awful lot as 100 devices concurrently, making it a powerful tool for reading galaxy clusters and one-of-a-type huge-scale systems inside the universe. It also can be used to have a take a look at the atmospheres of exoplanets, seeking out signs of existence or habitability.

The Mid-Infrared Instrument (MIRI) is a specifically touchy tool that is designed to have a have a look at the faintest and coldest items within the universe. It is mainly useful for analyzing the formation of stars and planets, and for trying to find the building blocks of existence in region. MIRI has the functionality to test in each imaging and spectroscopy modes, allowing it to examine a massive sort of devices and phenomena.

The Fine Guidance Sensor/Near InfraRed Imager and Slitless Spectrograph (FGS/NIRISS) is the fourth and very last tool onboard the JWST. It is designed to offer excessive-precision pointing and stability manipulate for the telescope, and to look at the atmospheres of exoplanets. FGS/NIRISS is a enormously bendy instrument that can be used for a big sort of clinical studies, from analyzing the dynamics of stars and galaxies to searching for dark depend and dark energy.

The technical capabilities of the JWST are not restrained to its devices. The telescope moreover has a number of modern-day-day capabilities a first rate manner to permit groundbreaking observations of the universe. For instance, its number one reflect is made from 18 hexagonal segments that can be adjusted for my part to accurate for distortions because of adjustments in temperature or gravitational forces. This guarantees that the telescope constantly offers you sharp and clean images, even in the most hard looking at conditions.

The JWST additionally has a very particular sunshield that is designed to protect the touchy devices from the warm temperature of the Sun. The sunshield is prepared the dimensions of a tennis court docket docket and is made from 5 layers of a completely unique cloth that displays and absorbs the warm temperature of the Sun. This ensures that the telescope stays cool, permitting it to make pretty touchy observations of the faintest gadgets inside the universe.

The observations an extraordinary way to be made with the resource of the JWST are expected to revolutionize our expertise of the universe. For example, the telescope can be able to check the earliest galaxies that long-established after the Big Bang, losing mild on the early universe and how it superior through the years. It also can be capable of have a observe the atmospheres of exoplanets in brilliant element, searching out the chemical signatures of water, methane, and precise molecules that could be indicative of existence.

The JWST can be able to take a look at the formation of stars and planets, and search for the constructing blocks of lifestyles in vicinity. It can also be able to study the properties of darkish rely and dark strength, mysterious additives of the universe that make up the bulk of its mass and electricity but whose nature stays unknown. By staring at the distribution of darkish rely in galaxy clusters and the cosmic microwave history radiation, the JWST will provide new insights into the character of these enigmatic additives.

Furthermore, the technical skills of the JWST will allow scientists to examine the universe in new and modern-day techniques. For example, the telescope's potential to observe in the infrared spectrum will allow it to look via dusty regions of the universe which is probably opaque to seen mild telescopes. This will allow us to take a look at the formation of stars and planetary systems in greater detail, similarly to the processes that power the evolution of galaxies.

In addition, the JWST's functionality to test more than one devices concurrently will permit us to look at large-scale systems within the universe, collectively with galaxy clusters and filaments, in more detail. By looking at those systems, we are able to find out about the large-scale shape of the universe and the strategies that drove its formation.

The technical talents of the JWST additionally present some annoying situations. One of the most vital worrying conditions is ensuring that the telescope remains strong and pointing correctly at some point of observations. To overcome this mission, the JWST is prepared with advanced gyroscopes and thrusters that could make small changes to the telescope's pointing and orientation.

Another undertaking is making sure that the telescope's sensitive gadgets stay loose from contamination. To cope with this, the telescope became assembled and tested in a smooth room environment, and it will probably be released into area on an Ariane five rocket, which has a established song file of reliability and safety.

Thus, the technical competencies of the JWST are not anything quick of modern-day. Its advanced devices and progressive functions will allow groundbreaking observations of the universe, losing mild on a number of the biggest mysteries in astronomy, which incorporates the origins of the universe, the character of darkish count and darkish energy, and the search for existence beyond our solar tool. While the traumatic conditions of running this shape of complicated observatory are giant, the medical rewards of the JWST are nice to be properly properly really worth the strive. We can assume a current era in astronomy, with the JWST as a imperative pillar of discovery and understanding.

Chapter 17: The Science Goals

The James Webb Space Telescope (JWST) is ready to notably exchange the take a look at of astronomy, allowing groundbreaking studies in quite a few areas. One of the maximum exciting additives of the JWST is its ability to have a study planetary structures, movie star formation, and exoplanets. In this bankruptcy, we are able to take a look at the technological understanding goals of the JWST in greater element, exploring how its abilties will allow us to check the ones phenomena in greater detail than ever earlier than.

Planetary System Exploration

The JWST's superior devices will permit us to have a take a look at planetary structures in extra detail than ever earlier than. One of the considerable technological records dreams of the telescope is to examine the formation of planets and their evolution over the years. By observing the infrared radiation emitted thru extra younger planets, the JWST is probably able to examine their atmospheres and stumble on the presence of gases which include water vapor, methane, and carbon dioxide. This will offer new insights into the conditions that exist in planetary systems and the manner they evolve over time.

The JWST will also be able to test the composition of asteroids and comets, which may be believed to be remnants from the formation of our solar device. By looking on the infrared radiation emitted by means of those objects, the telescope can be capable of decide their composition and provide new insights into the early facts of our sun device.

Star Formation

The JWST's ability to have a look at within the infrared spectrum may even permit us to study the method of big name formation in remarkable detail. By reading the infrared radiation emitted via protostars, the telescope may be capable of decide their temperature, mass, and chemical composition, supplying new insights into the formation of stars.

The JWST additionally may be able to have a look at the surroundings spherical greater youthful stars, which consist of the dust and fuel clouds that surround them. By searching at the infrared radiation emitted via using those objects, the telescope might be capable of decide their composition and provide new insights into the way of movie star formation.

Exoplanets

Exoplanets are planets that orbit stars apart from our sun, and the discovery of heaps of those planets in present day years has converted our information of the universe. The JWST could be capable of test the atmospheres of exoplanets, providing new insights into their composition and the situations that exist on their surfaces.

One of the essential element targets for the JWST can be the have a have a look at of exoplanets that are comparable in period and composition to Earth. By searching on the infrared radiation emitted by way of means of those planets, the telescope is probably able to stumble on the presence of gases which consist of oxygen, methane, and water vapor, which can be indicative of a liveable surroundings.

The JWST can also be capable of observe the atmospheres of fuel massive exoplanets, which consist of Jupiter and Saturn. By looking the infrared radiation emitted thru these planets, the telescope can be able to decide the composition of their atmospheres, supplying new insights into their formation and evolution.

Challenges

The JWST has numerous benefits, but it moreover has a number of troubles in its quest to take a look at planetary structures, superstar formation, and exoplanets. Making first-rate the telescope is everyday and pointed exactly in the course of observations is one of the most vital problems. This is crucial for learning dim items like exoplanets, which want extended publicity periods to be detected.

The safety of the telescope's sensitive device in competition to infection is each different trouble. In order to remedy this, the telescope have become put together and examined in a clean laboratory putting. It is probably despatched into orbit on the dependable and stable Ariane five rocket.

The Challenges of JWST: Launch and Deployment

The JWST represents one of the maximum ambitious area missions in facts, with the reason of redefine our knowledge of the universe. However, the achievement of the mission is based upon at the successful launch and deployment of the telescope, which gives some of worrying conditions. This financial ruin will therefore have a have a look at the traumatic conditions of launching and deploying the JWST in greater detail, exploring the measures taken to make sure the telescope's fulfillment.

The JWST is ready to launch in 2021, following a few years of making plans and development. The release will take vicinity aboard an Ariane five rocket, which has a tested tune file of reliability and protection. However, launching a spacecraft into vicinity is a complex way that requires specific planning and execution to make sure success.

Making effective the spacecraft can withstand the hard release conditions is one of the crucial troubles of the discharge method. High tiers of vibration, acceleration, and heat pressure might be felt thru the spacecraft in the course of liftoff. To combat this, the JWST turn out to be constructed to be as long lasting as possible, withstanding excessive levels of stress and vibration.

Another problem is making sure the spacecraft can launch on time, whilst you undergo in mind that delays can be high-priced and might have an impact on the task's final effects. The release manner has been meticulously deliberate and achieved, and backup measures had been installed region in case of unanticipated headaches, to reduce the hazard of delays.

Once the JWST reaches area, it's going to need to be deployed and prepared for observations. This gives each specific set of challenges, due to the fact the telescope want to be carefully located and calibrated to make certain that it capabilities effectively. The deployment device is mainly complex, due to the fact the telescope should be unfold out and assembled in area, a manner that has in no way been tried before.

To cope with this project, the JWST has been designed to be as modular as viable, with quite some its additives able to be separated and deployed independently. This will permit the telescope to be assembled step by step, with each component carefully positioned and examined before the subsequent is deployed.

Another venture of the deployment approach is making sure that the telescope's touchy devices stay loose from contamination. To cope with this, the telescope changed into assembled and examined in a smooth room environment, and it is going to be launched into location with a sunshield with the intention to protect it from contamination and thermal stress.

The sunshield is a key thing of the JWST, designed to guard the telescope from the acute warmth of the sun and hold it at a strong temperature. The sunshield is made from 5 layers of a specialised fabric called Kapton, which is robust and mild-weight, and gives amazing thermal insulation.

However, the deployment of the sunshield affords a pinnacle mission, as it want to be unfolded and positioned correctly in place. To deal with this, the sunshield has been designed to be as modular as viable, with every layer able to be deployed independently. This will permit the sunshield to be assembled regularly, with every layer carefully positioned and tested before the following is deployed.

Chapter 18: The Data Processing and Analysis

The James Webb Space Telescope (JWST) is designed to seize immoderate-choice pictures of the universe, permitting groundbreaking research in a number areas, from planetary system exploration to celebrity formation and exoplanets. However, the achievement of the challenge relies upon not simplest on the a achievement launch and deployment of the telescope, but furthermore at the accurate processing and evaluation of the information it collects. In this chapter, consequently, we will find out the records processing and analysis techniques and strategies used by the JWST challenge.

Data processing and evaluation are crucial additives of any clinical venture, as they permit researchers to extract full-size insights from the information gathered. The statistics accrued with the useful aid of the JWST can be massive and complicated, requiring current strategies and strategies to method and have a look at.

One of the primary stressful situations of processing and reading information from the JWST is the sheer quantity of records an awesome way to be accrued. The telescope's superior gadgets are capable of taking pictures pictures and spectra of items within the universe with tremendous detail, generating big quantities of facts that want to be processed and analyzed. To cope with this venture, the JWST institution has advanced a number of data processing and analysis techniques which might be specifically tailor-made to the particular characteristics of the telescope.

One of the essential factor statistics processing techniques used by the JWST is statistics discount. Data bargain includes taking the uncooked records accrued with the aid of the telescope and changing it into a more usable format. This gadget includes eliminating noise and artifacts from the records, calibrating the devices, and correcting for any mistakes brought at some stage in the statistics series technique. The resulting statistics is then geared up for evaluation.

Another principal records processing technique is statistics compression. Data compression includes reducing the scale of the statistics files with out losing important facts. This is vital for the JWST mission, because it will permit the transmission of huge portions of records decrease once more to Earth without overwhelming the available bandwidth.

The JWST crew has also evolved a number evaluation strategies which might be specially tailor-made to the kinds of information collected via the use of the telescope. For example, the telescope's Near-Infrared Camera (NIRCam) is able to shooting immoderate-choice photographs of the universe within the close to-infrared spectrum. To take a look at those pictures, the team has superior specialised picture processing techniques that allow them to pick out out and test devices within the photographs.

In addition to picture processing strategies, the JWST group has additionally evolved advanced spectral evaluation strategies. The telescope's Near-Infrared Spectrograph (NIRSpec) is able to capturing spectra of gadgets within the universe, that could offer valuable facts approximately their composition, temperature, and one in every of a type inclinations. The team has developed specialized techniques for studying the ones spectra, which contain comparing them to fashions of diagnosed spectra to discover the particular spectral capabilities of the item being studied.

The JWST organization has moreover superior sophisticated modeling techniques that permit them to simulate the universe based at the information gathered through way of the telescope. These models may be used to check hypotheses and make predictions approximately the conduct of gadgets inside the universe.

Another outstanding element of records processing and evaluation for the JWST venture is collaboration. The telescope can be operated by means of using way of a group of scientists and engineers from round the arena, each with their private areas of knowledge. To ensure that the facts collected thru the telescope is effectively processed and analyzed, the team has installation some of collaborative tools and systems that allow researchers to art work together and percent their findings.

The Future of Astronomy with JWST: Implications and Potential Discoveries

The James Webb Space Telescope's superior gadgets and abilities will permit researchers to have a examine the universe with extraordinary element and readability, putting in up new avenues for discovery and exploration. In this phase, we're able to have a take a look at the potential implications and discoveries that might be made with the JWST, and what the future of astronomy must appear to be as a quit result.

Planetary System Exploration

One of the important component areas of hobby for the JWST project is planetary machine exploration. The telescope's superior capabilities will permit researchers to observe the atmospheres of planets in our very very very own solar system, in addition to planets orbiting wonderful stars. By studying the composition and conduct of those planets, researchers want to advantage a better information of the way planetary structures shape and evolve.

One functionality discovery that is probably made with the JWST is the detection of biosignatures inside the atmospheres of exoplanets. Biosignatures are chemical signatures that advocate the presence of lifestyles on a planet, consisting of the presence of oxygen inside the surroundings. If biosignatures are detected on an exoplanet, it could be a strong indication that life exists on that planet. This discovery may also want to have profound implications for our statistics of the universe and our place in it.

Another discovery is the detection of water on planets in our very own sun gadget. The telescope's superior gadgets are able to detecting the faintest signs of water molecules, that could allow researchers to have a observe the distribution and behavior of water on planets like Mars and Europa. This statistics might be used to better apprehend the ability habitability of those planets, and could offer vital insights into the origins of life within the universe.

Star Formation

Another place of interest for the JWST mission is well-known character formation. The telescope's advanced abilities will permit researchers to look at the earliest stages of movie star formation, whilst clouds of gas and dirt crumble to form new stars. By studying this method in element, researchers desire to advantage a better understanding of approaches stars form and evolve, and the way they form the galaxies wherein they may be discovered.

One different discovery is the detection of protostars, which can be the earliest degrees of massive call formation. These gadgets are hard to test with gift telescopes, as they may be shrouded in dense clouds of fuel and dust. The JWST's advanced infrared abilities will permit researchers to penetrate the ones clouds and feature a look at the protostars in detail, providing precious insights into the early degrees of movie star formation.

Exoplanets

Perhaps the most thrilling vicinity of studies enabled thru the JWST challenge is the study of exoplanets - planets orbiting amazing stars. The telescope's superior talents will permit researchers to check the composition and behavior of those planets in notable detail, imparting precious insights into their origins, evolution, and capability habitability.

One functionality discovery that is probably made with the JWST is the detection of exoplanets with Earth-like atmospheres. These planets may want to have a comparable composition and conduct to Earth, making them strong applicants for the lifestyles of existence. The telescope's advanced devices are capable of detecting the faintest indicators of those atmospheres, allowing researchers to look at their composition and behavior in detail.

Another discovery is the detection of exomoons - moons orbiting exoplanets. These devices are hard to come across with cutting-edge telescopes, however the JWST's superior abilities might also need to permit researchers to find out the faint indicators of these items and have a look at them in detail. Exomoons might be crucial signs and symptoms of the habitability of exoplanets, as they will offer environments which can be conducive to the improvement of life.

JWST Collaborations and Partnerships: International Cooperation and Contributions

Collaboration and partnerships are key components of a fulfillment area exploration missions, and the James Webb Space Telescope (JWST) isn't any exception. The telescope has been advanced through a collaborative strive among NASA, the European Space Agency (ESA), and the Canadian Space Agency (CSA), and will include contributions from scientists and engineers from round the area. In this segment, we are able to observe the global cooperation and contributions which have been instrumental within the improvement and achievement of the JWST undertaking.

NASA, ESA, and CSA Collaboration

The improvement of the JWST has been a collaborative attempt between NASA, ESA, and CSA. Each business enterprise has made large contributions to the challenge, with NASA liable for the improvement and creation of the telescope, ESA presenting the Ariane 5 rocket an fantastic way to launch the telescope into space, and CSA imparting the Fine Guidance Sensor (FGS) to be able to allow the telescope to as it have to be component and stabilize itself in region.

NASA has been the lead organization inside the improvement of the JWST, overseeing the format, introduction, and finding out of the telescope. The organization has contributed the majority of the investment for the venture, and has labored closely with its global partners to make sure the success of the undertaking.

ESA has been responsible for the development of the Ariane five rocket so as to release the telescope into location. The commercial employer corporation has additionally contributed severa of the telescope's key devices, which includes the Near-Infrared Spectrograph (NIRSpec) and the Mid-Infrared Instrument (MIRI).

CSA has contributed the FGS, that's a critical element of the telescope's pointing and stabilization gadget. The FGS will permit the telescope to because it must be element at its goal gadgets, ensuring that the pics and records gathered are of the high-quality great.

International Contributions to the JWST Mission

In addition to the contributions made by way of way of the 3 lead businesses, the JWST project has worried good sized worldwide collaboration and contributions. Scientists and engineers from spherical the sector have completed crucial roles within the development and locating out of the telescope's devices and systems, and could keep to play a key position inside the evaluation and interpretation of the statistics amassed through using the telescope.

One of the most good sized worldwide contributions to the JWST challenge has been the development and creation of the telescope's gadgets. In addition to the devices advanced via way of ESA, numerous one-of-a-kind worldwide partners have contributed gadgets to the undertaking, consisting of the Canadian Space Agency (CSA), the University of Arizona, and the Max Planck Institute for Astronomy in Germany.

The CSA's contribution of the FGS is sincerely one instance of the commercial enterprise company's huge contribution to the JWST assignment. The FGS is a essential issue of the telescope's pointing and stabilization machine, and could permit the telescope to effectively aspect at its aim gadgets.

The University of Arizona has additionally made giant contributions to the JWST venture, developing severa key additives of the telescope's gadgets, along with the Near-Infrared Camera (NIRCam) and the Near-Infrared Imager and Slitless Spectrograph (NIRISS). These gadgets will permit the telescope to take a look at the universe in wonderful detail, imparting valuable insights into the origins and evolution of galaxies, stars, and planets.

The Max Planck Institute for Astronomy has also completed a big position within the improvement of the JWST venture, contributing the Mid-Infrared Instrument (MIRI). This tool will permit the telescope to study the universe inside the mid-infrared vicinity of the electromagnetic spectrum, imparting treasured insights into the behavior of dust and fuel within the universe.

International collaboration and partnerships also can be essential in the assessment and interpretation of the statistics accrued thru the JWST. Scientists from spherical the world will artwork collectively to investigate the facts accrued through the use of the telescope, sharing their records and insights to advantage a better information of the universe and its many mysteries.

Chapter 19: The Role of Citizen Science

Citizen generation has achieved an an increasing number of crucial function in astronomy studies in latest years, and the James Webb Space Telescope (JWST) assignment has had a truthful percentage of it. Citizen technological expertise entails attractive individuals of the general public in scientific studies, allowing them to contribute to crucial discoveries and advantage a deeper information of the universe. In this economic break, we will speak the location of citizen generation within the success of the JWST challenge and the functionality advantages it is able to provide.

What is Citizen Science?

Citizen technology is a collaborative method to medical studies that involves attractive people of the general public inside the scientific way. Citizen technological know-how tasks can take many office paintings, from collecting records in the place to analyzing data amassed through professional scientists. In latest years, citizen technological know-how has become an increasing number of famous in astronomy research, with severa duties appealing individuals of the general public within the look for exoplanets, the have a examine of galaxy morphology, and the elegance of galaxies.

Benefits of Citizen Science in Astronomy

Citizen technological facts has some of capability advantages for astronomy studies. By engaging contributors of the general public in scientific studies, citizen technology tasks can growth the scope of research and data series, contemplating extra entire and accurate effects. Citizen science projects also can help to democratize technological know-how, making it handy to a wider target market and provoking more public engagement with generation.

In the context of the JWST venture, citizen technological know-how has the capacity to contribute drastically to the fulfillment of the challenge. By engaging people of the overall public within the search for exoplanets and the take a look at of galaxy morphology, citizen technology initiatives can help to growth the scope of the JWST's clinical studies, offering treasured facts and insights that expert scientists may not be capable of collect on their private.

Examples of Citizen Science Projects in Astronomy

There are severa examples of citizen era responsibilities in astronomy, many of that have already made huge contributions to our know-how of the universe. One instance is the Zooniverse challenge, which permits individuals of the overall public to classify galaxies primarily based mostly on their morphology. The mission has already categorized tens of hundreds of thousands of galaxies, supplying valuable insights into the distribution and evolution of galaxies within the universe.

Another example is the Planet Hunters undertaking, which lets in members of the general public to look for exoplanets through studying facts from the Kepler vicinity telescope. The venture has already identified numerous ability exoplanets, severa of that have been showed via way of professional astronomers.

The Future of Citizen Science in JWST's Success

Citizen studies is anticipated to grow to be greater important to the success of the JWST assignment as it movements forward. The telescope is the right device for citizen era duties due to its unequalled talents and capacity to look the cosmos in awesome detail. This lets in individuals of the overall public to contribute to widespread clinical enhancements and deepen their expertise of the universe.

The look at of exoplanets is one feasible use of citizen technological statistics in the JWST assignment. The telescope is the suitable tool for citizen scientists to have a look at statistics and look for new exoplanets considering the reality that it is able to locate and take a look at exoplanets in notable detail. Citizen technology obligations with an exoplanet emphasis have the functionality to discover new planets and make stronger our expertise of the distribution and behavior of exoplanets within the cosmos.

The investigation of galaxy morphology is however every other viable field for citizen research within the JWST assignment. Due of the telescope's unmatched capacity to check galaxies in element, citizen scientists may use it to take a look at and categorize galaxies in keeping with their morphology. Citizen technology initiatives dedicated to the evaluation of galaxy morphology have the potential to discover novel types of galaxies and beef up our information of the cosmological improvement of galaxies.

The Social and Ethical Implications of JWST's Discoveries

As with any maximum important clinical assignment, the potential discoveries from

JWST's observations encompass social and moral implications that want to be taken into consideration. In this section, we are capable of discover some of the important factor social and moral implications of JWST's discoveries and their functionality effect on society.

The most exciting opportunities of JWST is its capacity to have a have a examine the origins of the universe, together with the formation of the primary galaxies, stars, and black holes. These observations will assist us to better apprehend our location inside the universe and our origins, but they will moreover mission some of our lengthy-held ideals approximately the universe and our area inside it. For example, JWST's observations also can moreover offer evidence for the lifestyles of a couple of universes, which can venture the concept that we live in a completely specific and unique universe.

JWST's observations of exoplanets moreover have social and moral implications. The discovery of doubtlessly habitable planets outside of our sun device ought to have profound implications for the look for extraterrestrial existence and our statistics of the individuality of life on Earth. If existence is placed on some different planet, it would boom questions on our dating with that lifestyles and the way we ought to interact with it.

The discovery of exoplanets moreover has implications for our know-how of the possibility of human settlement on precise planets. JWST's observations of the atmospheres of exoplanets may also moreover need to assist us to better apprehend the conditions critical for lifestyles and the capability habitability of various planets. This may want to inform future missions to find out and colonize other planets, however additionally increases questions on the ethics of terraforming different planets and the effect of human colonization on the ecosystems of numerous planets.

Another social and moral implication of JWST's discoveries is the effect on our information of faith and spirituality. For many people, the concept of lifestyles on exclusive planets or a couple of universes disturbing situations their spiritual beliefs and will growth questions about the character of God and the cause of human lifestyles. It is critical to understand and respect the range of religious ideals and perspectives on the ones troubles.

The functionality discoveries from JWST additionally have implications for public insurance and funding. The discovery of probable liveable planets or evidence of extraterrestrial existence need to have a extensive impact on public opinion and coverage alternatives related to area exploration and investment for scientific research. It is crucial to don't forget the capacity impact of JWST's discoveries on public opinion and to ensure that selections approximately funding and coverage are primarily based definitely totally on clinical proof and no longer inspired via way of political or ideological issues.

Finally, the discoveries from JWST also have implications for schooling and outreach. JWST's discoveries have the capacity to encourage destiny generations of scientists and to boom public focus and understanding of the universe and our vicinity inner it. It is essential to put money into schooling and outreach applications to ensure that the general public is informed and engaged within the scientific discoveries from JWST.

Reflections on JWST's Contributions to Astronomy and its Legacy

It is critical to remember the James Webb Space Telescope's (JWST) feasible contributions to astronomy and its legacy as it prepares to convert our information of the cosmos. JWST is expected to discover novel phenomena and offer higher insights into the universe' beginnings and improvement thanks to its contemporary-day-day device. In this bankruptcy, we're going to have a take a look at some of the medical advances and discoveries that the JWST may moreover make as well as how it might affect our facts of the cosmos in the long run.

Investigating the beginnings and development of galaxies is one of the crucial dreams of JWST. The telescope will allow us to see galaxies which may be older and further away than the ones that would currently be located with present day telescopes due to its improved sensitivity and resolution. This will offer moderate at the creation of the earliest stars and black holes in addition to the development of galaxies and their genesis. The reionization of the cosmos and the tendencies of the intergalactic medium will every be positioned with the aid of JWST's studies of excessive-redshift galaxies, giving notion on the time when the universe modified from a independent to an ionized usa.

JWST will substantially beautify our statistics of ways stars and planetary systems rise up and increase. The telescope will help in offering crucial answers to critical queries much like the beginnings of lifestyles and the potential existence of liveable planets beyond of our sun tool through studying the strategies of famous man or woman and planet formation in extraordinary element. With the help of JWST, we're able to be able to take a look at the atmospheres of exoplanets and determine if any of them contain water or other biomarkers that may be signs and symptoms of alien existence. These findings ought to significantly have an effect on how we see the cosmos and the manner we search for extraterrestrial existence.

JWST's present day device will permit a large sort of astronomical take a look at further to those important clinical desires. Its immoderate-choice imaging abilties, for instance, will permit in-depth research of the dynamics and form of galaxies, on the identical time as its spectroscopic talents will offer facts about the chemical makeup and physical traits of celestial devices. The JWST's infrared talents also can allow studies into the synthesis of complex molecules and the interstellar medium, on the way to provide mild at the universe's chemical history.

Furthermore, the information amassed by way of JWST may be a beneficial tool for astronomers inside the destiny years. The telescope is anticipated to provide an high-quality amount of exceptional records if you want to be finished for many years to move back because of its extended strolling lifespan and complicated gadget. The facts may be made reachable to the overall public, giving citizen scientists and amateur astronomers the chance to make contributions to astronomical research and discoveries.

www.ingramcontent.com/pod-product-compliance
Lightning Source LLC
Chambersburg PA
CBHW071443080526
44587CB00014B/1968